Architectural Acoustics
Design Guide

Architectural Acoustics
Design Guide

Acentech
James Cowan, Senior Consultant

McGraw-Hill
New York San Francisco Washington, D.C. Auckland Bogotá
Caracas Lisbon London Madrid Mexico City Milan
Montreal New Delhi San Juan Singapore
Sydney Tokyo Toronto

McGraw-Hill

A Division of The *McGraw·Hill* Companies

1 2 3 4 5 6 7 8 9 0   DOC/DOC   9 0 9 8 7 6 5 4 3 2 1 0 9

ISBN 0-07-135938-9

The sponsoring editor for this book was Wendy Lochner, the editing supervisor was Virginia Carroll, and the production supervisor was Pamela Pelton. It was set in Matt Antique by North Market Street Graphics.

Printed and bound by R. R. Donnelley & Sons Company.

 This book is printed on recycled, paper containing a minimum of 50% recycled, de-inked fiber.

McGraw-Hill books are available at special quantity discounts to use as premiums and sales promotions, or for use in corporate training programs. For more information, please write to the Director of Special Sales, McGraw-Hill, Two Penn Plaza, New York, NY 10121-2298. Or contact your local bookstore.

To Phil Praino for all your contributions, both personal and professional, to Acentech and its staff. You have been a role model of strength and integrity to all of us.

Contents

Foreword

Acoustics has been the least understood ingredient of building design since the early Roman and Greek periods. Every person has horror stories about headaches following evening meals in noisy restaurants. Or the impossibility of understanding lectures in school auditoriums. Or the acoustically transparent walls in budget motels or even in quality apartment houses. And who has not heard of the impossible acoustics in some hall supposedly designed for concert music? My reading finds that this book is effectively dedicated to the guidance of architects and designers toward the satisfactory creation of quality acoustical environments—with indications where professional help ought to be sought.

James Cowan has done a masterful job of clarifying the basics of good acoustical design by two means. The first is a recent release of a multimedia CD-ROM, titled *Architectural Acoustics* (McGraw-Hill, 1999), in which he takes the architectural designer on an acoustical tour that melds aural concepts with visual realizations. The second is this book, which presents design tables and structures, along with drawings and photographs of projects within his firm's experience.

I especially appreciated the book's organization. The opening chapters present an easily understandable overview of the fundamentals of architectural acoustics and the basic ways that sound can be controlled in buildings and outdoors. The reader will learn how acoustical waves behave when they are reflected, are bent around partial partitions or outdoor walls, or encounter irregular surfaces. Explained are decibels, which in some ways are analogous to degrees on a thermometer, and how they are measured. Valuable tables are incorporated that present the effectiveness of various materials in reducing sound levels in a room, or in blocking the transmission of noise from one room to another through walls or ceilings/floors.

In the middle chapters, basic charts are presented that, together, assemble a kit of design tools requisite to the identification and control of specific acoustical phenomena. Interspersed with the charts are examples of spaces in which they are effectively utilized. I was taken especially by the way that the charts correlate design solutions with the primary purposes of rooms and how the interwoven case studies are illustrated by excellent photographs and drawings. The reader needs no mathematical skills. The book ends with a nonmathematical glossary of the terms used in the text and a few basic equations for commonly cited acoustical quantities.

I feel confident that a careful study of this book and supplementary immersion in the visual/voice tour of the aforementioned CD-ROM will constitute a highly rewarding introduction to the needs of architects and designers who deal with architectural projects that involve acoustical considerations.

Leo L. Beranek
Author of *Concert and Opera Halls: How They Sound*,
cofounder of Bolt Beranek & Newman, Inc., Founding
President, Institute of Noise Control Engineering
Past President, Acoustical Society of America

Acknowledgments

This book would not have been possible without the support of Acentech Incorporated or their staff. The case studies that comprise most of this book were projects that were managed and performed by Acentech personnel, including Carl Rosenberg, Rein Pirn, Tom Horrall, Chris Savereid, Frank Iacovino, Paul Burge, Parker Hirtle, Bob Berens, Doug Sturz, Larry Philbrick, Eric Ungar, and Kurt Milligan. Contributions to the book were also made by Paul Remington, Bob Jones, and Hal Amick of BBN. Figures were generated by Robyn Spencer and Phil Praino. Most of the photography was performed by Joan McQuaid, unless otherwise noted in the figures. Assisting us with our photographical pursuits were Bruce Spena at the Berklee College of Music, Peter Waldron at the Rogers Center for the Arts, Fran Pedone at the Worcester Art Museum, Curtis Eckelkamp and Norman Noel at the Lowell Memorial Auditorium, and Rick Ryerson at Harley-Davidson's offices. The editorial comments by David Harris and Leo Beranek were greatly appreciated in finalizing this book.

Introduction

In the summer of 1999, McGraw-Hill released a multi-media CD-ROM set entitled *Architectural Acoustics*. That product was designed to be an educational tool for architects and designers from which they could hear the principles of acoustics described. This book, written by the same author as the *Architectural Acoustics* CD set, has its basis in the principles described in the multimedia CDs, but goes one step further to give examples of actual designs that successfully incorporate those principles.

From the theaters of ancient Greece to those of the twenty-first century, architectural acoustics has been a key consideration in design. Only within the past century, however, have we been able to scientifically understand and predict the behavior of sound both indoors and outdoors. It is through this understanding that acoustics has evolved from a black art into an established field of engineering.

This book explains the basic principles of acoustics and offers examples of how they have been effectively incorporated into current architectural design. A unique aspect of this book is that members of the author's acoustical consulting firm (including some from their former parent firm, Bolt Beranek & Newman, Inc.) designed the acoustic facets of all buildings referenced in its case studies. Such intimate involvement in each project affords the reader a complete picture of how the acoustics was considered. The following is an overview of the organization and contents of this book.

After summarizing the acoustic principles discussed in the CD set in Part 1, Part 2 of this book is divided into sections that address specific acoustic principles, each accompanied by its own checklist of design tools and descriptions of designs that exemplify the advantages of using those design principles.

Chapter 1 covers the basic points that need to be understood about acoustics before any other discussions occur. Each point in this chapter builds on information supplied in each successive section. The basics of sound generation and propagation form the foundation. Then frequency, wavelength, and speed of sound are used to describe the properties of sound waves. A discussion on our hearing process is included next. Any instructional material on acoustics cannot be complete without allowing the reader to understand how we perceive sound, or how we hear.

The next section deals with the ways in which sound waves change their direction of travel after encountering a change in medium. These can be separated into the general categories of reflection, refraction, diffraction, and diffusion. These principles will make any later discussions about indoor and outdoor sound travel understandable. The final topic in this discussion of basic acoustics is how we describe sound levels in terms of decibels. This is one of the most misunderstood topics in the field, but can be appreciated only after comprehending the material leading up to it in this chapter.

Chapter 2 discusses indoor and outdoor sound control. When considering materials for acoustics, the principles of indoor sound control can be divided into two categories: absorption and insulation. The absorption of materials dictates their ability to control sound *within a room* while the insulation properties of materials dictate their ability to control sound *between rooms*. The redirection of sound is also covered to emphasize that the control of sound does not always imply the reduction of sound energy. In some cases, we do not want to reduce the sound, but only to control where it is going. These discussions naturally lead to explanations of effective noise reduction methods.

The number of practical options for outdoor sound control is more limited than that for indoor sound control. These options include buffer zones, noise barriers, and masking systems.

The rest of the book (Chapters 3 through 7) offers charts listing the design tools to be considered to control specific acoustic phenomena. Following the design tool charts, there are examples of buildings where these design tools have been used to effectively consider these phenomena. Chapter 3 shows the principles of room acoustics and reverberation control, using room shapes to effectively control sound, and interior noise control design. Chapter 4, "Sound Privacy/ Isolation," covers design tools and examples of low-frequency absorption and transmission loss, noise reduction and privacy, sound masking and background noise control, and acoustic issues common to multiple-unit residential and commercial buildings. Chapter 5 is dedicated to the unique sound privacy challenges of open-plan offices. Chapter 6 addresses multipurpose facilities, and Chapter 7 covers outdoor sound control.

This book deals only with the practical design aspects of spaces rather than getting bogged down with mathematics and jargon. There are no mathematical equations in the main body of this book. There are some mathematical descriptions, but only to get the appropriate points across. A few essential mathematical equations are included in Appendix A (to define acoustical terms), but those interested in more detailed mathematical equations and calculations of quantities should consult the many other fine books on acoustics that have already been written. For those interested in further exploration of commonly requested information from architects, a Technical Addendum has been included, containing articles covering these topics in more detail than is found in the text.

Architectural Acoustics
Design Guide

Overview of the Fundamentals of Architectural Acoustics

CHAPTER 1

Basic Acoustics

Acoustics

Acoustics is the science of sound. We interpret sound through our sense of hearing. Anything that is interpreted by the senses is open to subjectivity in terms of likes and dislikes. This subjective interpretation of sound not only defines the differences between music and noise, but also dictates the quality of communication within a space. People often think of acoustics as a narrow, esoteric field that has little practical application short of designing concert halls. But the field has many practical branches, including noise control, psychoacoustics (the psychological effects of sound on people), physiological acoustics (the physical effects of sound on people), bioacoustics (the use of sound waves in medical diagnostics), and the subject of this book—architectural acoustics.

Architectural acoustics deals with sound in the built environment. In many ways, especially with music, this field is controversial because of the wide variation in personal tastes. In many other ways, however, the field of architectural acoustics deals with accepted scientific principles. This book will describe these criteria, in terms of the principles themselves and in terms of practical examples. This book is not a treatise on concert hall design or the design of spaces requiring subjective evaluation. There are many quality books that have already been published on this subject. See a partial listing of these books included in Appendix B.

The most logical place to begin our discussion of architectural acoustics is at the source of the sound wave. We will now discuss general classes of sound sources, along with their most general characteristics.

Sound Generation

Sound is generated when pressure oscillations are generated in an elastic medium at rates that are detectable by a hearing mechanism. For simplicity and practicality, the sound waves described in this book are in the range detectable by most humans. Sound generation is a physical phenomenon while noise is a subjective interpretation of sound. Therefore, if a tree falls in the woods and no one is nearby to hear it, the tree has generated a sound but not noise (putting to rest a long-debated argument).

If the source of pressure oscillation is stationary (relative to a stationary observer) and physically small compared to your distance away from that source, we call it a *point source*. If this source is oscillating at a constant rate, it generates a *pure tone* and the source can be described in terms of a single *frequency,* or rate of oscillation. This frequency is usually described in terms of units of cycles (of oscillations) per second, also labeled as *hertz* (abbreviated Hz), named after the German physicist Heinrich Hertz, who is credited with discovering electromagnetic radiation waves.

If one were to visualize the generation of a pure tone, Figure 1.1 would be most instructive. The constant rate of oscillation with the passage of time translates to the sine wave shown in Figure 1.1. One complete cycle of oscillation is shown ending at the point at which the sine wave begins to repeat its pattern. Pure tones rarely exist in nature since most sound comprises contributions from many audible frequencies.

It is generally accepted that humans can hear frequencies between 20 and 20,000 Hz. Within this frequency range, we are most sensitive to sounds having frequency components between 500 and 4000 Hz. This, by no coincidence, also corresponds to the dominant frequency range generated by the

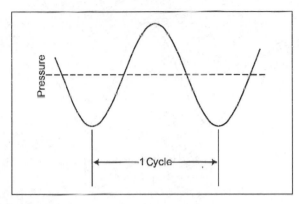

Figure 1.1 Pressure fluctuations in time for a pure tone sound wave.

human voice. Although most of us can still detect low pitches between 20 and 500 Hz and high pitches between 4000 and 20,000 Hz, our hearing mechanisms are less sensitive to these sounds. We will explore this further in our discussion of the hearing process later in this chapter.

Frequencies below 20 Hz are known as *infrasonic*. These frequencies, although not audible to most people, can be felt as vibrations. This is due to the fact that our internal organs resonate at frequencies between 5 and 15 Hz. Each physical entity has a resonance frequency associated with it, depending on its density. Exposure to sounds near the resonance frequency of a material causes it to vibrate more than it would when exposed to other frequencies. In extreme cases, high levels of infrasound can interfere with the healthy operation of our internal organs. It is therefore not advisable to frequent areas where you can feel your chest or abdomen vibrating.

Frequencies above 20,000 Hz are known as *ultrasonic*. These frequencies, for reasons to be mentioned in the upcoming discussion on wavelength, can be focused into narrow beams that can be useful for applications ranging from medical diagnostics (to view internal organs or fetuses) to drilling and cleaning teeth.

Figure 1.1 shows a single-frequency sound wave traveling with time. If the horizontal axis in this plot were replaced with distance, we would be viewing the physical

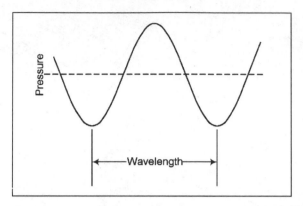

Figure 1.2   Pressure fluctuations in space for a pure tone sound wave.

length of the sound wave. Since a complete cycle of oscillation corresponds to the time it takes for the wave to repeat its pattern, the *wavelength* is the distance over which the wave travels before it begins to repeat itself. This is shown in Figure 1.2.

The wavelength of sound is inversely proportional to its frequency. In other words, the length of the wave increases as frequency decreases, and the length of the wave shortens as frequency increases. The third term in the equation (listed in Appendix A) that describes this phenomenon is the speed of sound. This is the speed at which the sound wave travels. The speed of sound depends on the density of the medium through which the sound wave is traveling (e.g., air, water, or solid materials) and the temperature. In air, the speed of sound varies between roughly 1050 (at 0°F) and 1150 (at 90°F) ft/sec (320 to 350 m/sec). At 70°F, the speed of sound is approximately 1128 ft/sec (344 m/sec).

Table 1.1 shows the relationship between frequency and wavelength over our audible frequency range at a temperature of 70°F. This table shows the range of wavelengths

Table 1.1   Relationship Between Frequency and Wavelength over the Audible Range

Frequency (Hz)	Wavelength
20	56 ft/17 m
50	23 ft/7 m
100	11 ft/3 m
500	2 ft/0.7 m
1,000	1 ft/0.3 m
5,000	0.2 ft/0.07 m
10,000	0.1 ft/0.03 m
20,000	0.06 ft/0.02 m

associated with our hearing sensitivity, from more than 50 ft (15 m) at 20 Hz to less than an inch (0.025 m) at 20,000 Hz. These values become important when attempting to control sound. This is because the wavelength must be less than the dimensions of a partition for that partition to effectively control the sound. This is true for reflection as well as noise reduction.

Sound generated from a point source radiates sound energy equally in all directions. This radiation pattern resembles a sphere, centered around the source. If the source were a train or a steady stream of traffic on a highway, it would not be appropriate to model the source as a point. It would be more appropriate to model it as a line, having a radiation pattern resembling a cylinder (or half of a cylinder if the source is on the ground). This type of sound source is known as a *line source*. Quantifying sound decay rates for point and line source travel will be explored at the end of this chapter in the section on the decibel.

The Hearing Process

Since acoustics has meaning only through our interpretation of sound, it is helpful to understand the human hearing mechanism to fully appreciate the science of acoustics. Figure 1.3 shows a cross section of the human hearing mechanism, divided into its most commonly labeled sections of the outer, middle, and inner ears.

The function of the outer ear is to funnel sound waves into the rest of the hearing mechanism. Without a *pinna* on the outside of our heads, we would not be able to hear most of the sounds around us. The picture of the man with an inverted horn held up to his ear accentuates this point. The ear canal funnels the sound to the *eardrum,* also known as the *tympanic membrane* in medical circles. The eardrum is the first location in the hearing mechanism where sound energy becomes converted into another form of energy in its

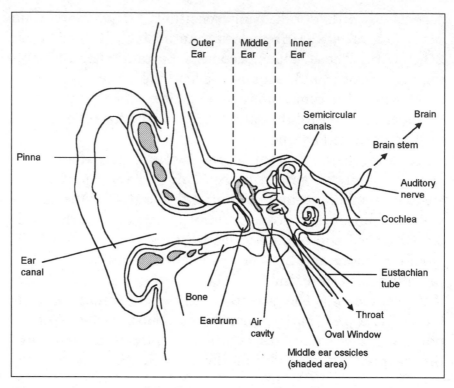

Figure 1.3 Cross section of the human hearing mechanism.

trip to the interpretation center of the brain. The eardrum also represents the limit of the outer ear.

The outer ear, by itself, is a cylinder that is open to the outside on one end (at the pinna) and closed at the other end (at the eardrum). This is similar to an organ pipe, tuned to a specific frequency by its dimensions. The typical human ear canal is roughly 1 to 1½ in (3 cm) in length. If this were an organ pipe of the same length, it would be tuned to frequencies in the 2000- to 3000-Hz range. We are therefore most sensitive to frequencies in this range, and our outer ears actually amplify sounds in that frequency range. There are advantages and disadvantages to this occurrence. In terms of advantages, 2000 to 3000 Hz lies in the upper end of the frequencies generated by human speech. These are the dominant frequencies of our consonant sounds, which assist us in communication with each other. The unfortunate part of this is that we tend to lose our hearing sensitivities in this frequency range first, thereby making communication more difficult. Hearing loss caused by exposure to high levels of sound causes the most profound loss of sensitivity in this frequency range.

Continuing on the path of a sound wave into the middle ear, the eardrum is set into vibration by the sound that is incident upon it. These vibrations are carried along the three small bones (called *ossicles*) in an air space comprising the middle ear. The *hammer, anvil,* and *stirrup* (known in medical circles as the *malleus, incus,* and *stapes,* respectively) carry the eardrum's vibrations to the oval window, which is the entrance to the inner ear. There are a few points worth noting about the functions of the middle ear. For one, these three bones serve as a transformer to adjust the volume of sound to one that is appropriate for the inner ear organs. Also, if sound levels are high, muscles that connect these bones cause them to separate, which in turn reduces the intensity of sound channeled to the inner ear. This reflex is not effective, however, for impulsive sounds since these types of sound events occur at a faster rate than the rate to which the protective mechanism can respond.

Since the air cavity of the middle ear is usually sealed from the outside world, changes in pressure outside our bodies are not experienced in the middle ear until that seal is broken. This imbalance in pressure is why changes in altitude (and, accordingly, atmospheric pressure) cause a pressure sensation behind our eardrums. The only connection between the middle ear and the outside world is the *eustachian tube* (named for the sixteenth-century Italian anatomist, Bartolomeo Eustachio), which runs from the middle ear to the throat. The sealed section is opened upon swallowing or yawning, actions that normally alleviate this sense of pressure in the middle ear.

Once the sound wave's energy reaches the oval window, it causes the oval window to vibrate. This, in turn, generates waves (similar to those on an ocean) in the fluid that fills the spiral-shaped apparatus known as the *cochlea*. The cochlea is lined with tiny, hair-sized cells that wave in the moving fluid. As these hair cells wave, they convert this mechanical energy into electrical energy and send these electrical signals to the auditory nerve. The auditory nerve then sends the electrical signals from all hair cells to the brain, where they are processed to interpret sounds. This entire hearing process takes only a fraction of a millisecond to occur.*

Changes in the Direction of Sound Travel

Sound waves change their direction of travel through four categories of phenomena: *reflection, refraction, diffraction,* and *diffusion*. These phenomena can occur when changes occur in a sound wave's medium of travel. These physical principles are the same as those that occur in the optical world with light. The principal difference between light and sound is the frequency range. Our visible frequency range is 1.6 to 2.8 billion Hz, while our audible frequency range is 20 to 20,000 Hz.

Reflection

When a sound wave encounters a sharp discontinuity in the density of a medium, some of its energy is reflected, as is illustrated in Figure 1.4. Reflected sound energy follows the laws of optics, with the simplest visualization analogy being a mirror. Just as light bounces off a mirror at the same angle as its angle of incidence, sound waves have equal angles of incidence and reflection. Reflective surfaces are typically smooth and hard.

*For more detailed information on the hearing mechanism and causes for hearing loss, see J. P. Cowan, *Handbook of Environmental Acoustics* (New York: John Wiley & Sons, 1997).

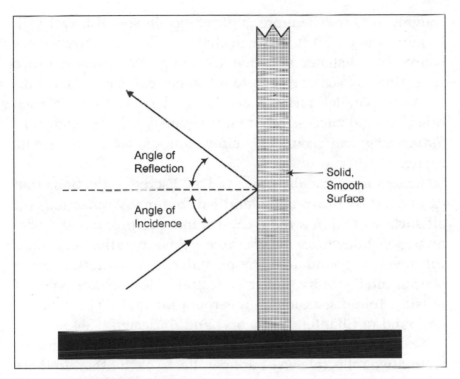

Figure 1.4 Reflection of a sound wave.

A few common room acoustics problems caused by reflection are echoes and room resonance. Echoes are caused by the limitations of our hearing mechanisms in processing sounds. When the difference in arrival times between two sounds is less than 60 milliseconds, we hear the combination of the two sounds as a single sound. However, when this difference exceeds 60 milliseconds, we hear the two distinct sounds. When these two sounds are generated from the same source, this effect (which we call an *echo*) can cause difficulty in understanding speech, especially when arrival times differ by more than 100 milliseconds. These kinds of delays occur when a person hears a sound coming directly from a source and one coming from a reflecting surface. Given that the speed of sound in air is roughly 1000 ft/sec, the 100-sec (or 0.1-sec) delay translates to a distance of roughly 100 ft. Therefore, a difference in sound travel path of more than 100 ft between that traveling directly from a source to a listener and that traveling from a source to a reflecting surface and then to a listener can cause an echo.

When parallel surfaces are tall and fairly close to each other, a rapid succession of midfrequency echoes, known as *flutter echo,* can occur. This effect sounds like birds or bats flying.

Echoes are normally perceived as discrete reflections that can be clearly heard and identified, but many reflections off all surfaces within a room can combine to produce the phenomenon known as *reverberation*. Reverberation can raise the level of sound in a room and can also detract from speech intelligibility, but it is desirable for certain types of music. More discussion on reverberation and its control can be found in Chapters 2 (theory) and 4 (design).

Room resonance occurs at specific frequencies in rooms where two reflective walls are parallel to each other. In these cases, whole-number multiples of specific half-wavelengths will fit between the two walls. Since their surfaces reflect the sound, their mirror images bounce off each wall to set up a stationary pressure pattern in a room. This phenomenon is called a single-dimensional (or axial) *standing wave* and it is the simplest form of room resonance. Standing waves can become more complex in two and three dimensions, where they are known as *tangential* and *oblique* modes, respectively. Figure 1.5 shows the pressure pattern of an axial standing wave for a room having parallel reflective surfaces.

A problem with standing waves for design purposes is their generation of an uneven sound distribution in a room. Some areas will have high levels of sound (because the standing wave is reinforcing the pressure at those locations) and some areas will have low levels of sound (because the standing wave is canceling much of the pressure at those locations).

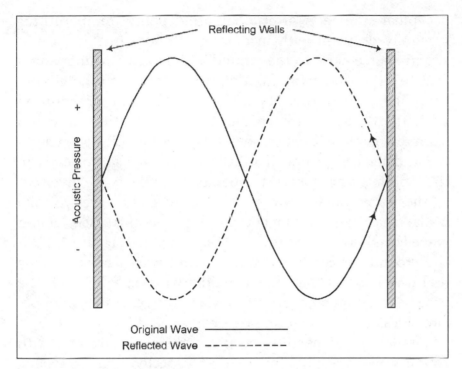

Figure 1.6 Generation of a standing wave.

Refraction

Just as light bends as it travels through a prism, the direction of sound is altered when sound waves encounter changes in medium conditions that are not extreme enough to cause reflection, but are enough to change the speed of sound. In addition to the speed of sound changing for different materials or media, the speed of sound changes with changes in temperature within the same medium. This variation in sound travel direction, caused by variations in the speed of sound, is known as *refraction*.

Sound refracts through outdoor areas where temperature changes. Because the speed of sound is faster in warmer air, sound waves bend toward cooler temperatures. This can be illustrated through two extreme examples. On a typical summer's afternoon, atmospheric temperatures generally decrease with increasing altitude. In this case, a sound source close to the ground will generate sound waves that bend up and away from distant listeners on the ground (see Figure 1.6). This effect does not become pronounced until the distances are greater than 200 ft from a sound source. Therefore, a sound source may be visible yet not audible at distances greater than 200 ft. Late at night or early in the morning, atmospheric temperature gradients are inverted. At these times, temperatures close to the ground are typically cooler than those at higher altitudes. In this case, sound waves bend downward toward the ground (see Figure 1.7). If the ground surface is reflective, sound waves bounce along and travel farther than one may otherwise expect. This is the case near a calm body of water, where conversations at opposite sides of lakes can often be clearly heard.

Similar sound bending occurs with wind currents, with sound waves traveling farther than expected when traveling with the wind and generating shadow zones (as are shown in Figure 1.6) when sound waves are traveling against the wind.

Diffraction

The principle of *diffraction* limits the sound reduction effectiveness of any open-plan office partition or outdoor

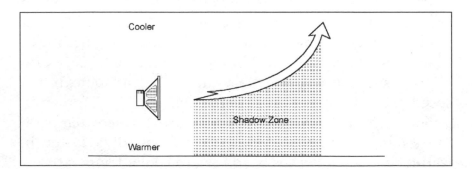

Figure 1.6 Refraction of sound as temperatures decrease with altitude.

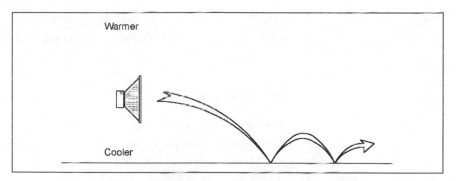

Figure 1.7 Refraction of sound as temperatures increase with altitude.

noise barrier. Sound waves bend around and over these types of walls, independent of their material, to impose this limit. Figure 1.8 shows this principle in action, creating (as with light) what is known as a *shadow zone* after the line of sight is broken between a sound source and a listener. The specific limits imposed by diffraction are described in Chapter 2 in the section on noise reduction methods.

Diffusion

When a sound wave reflects off a convex or uneven surface, its energy is spread evenly rather than being limited to a discrete reflection. This phenomenon, known as *diffusion* (see Figure 1.9) is the acoustic equivalent to the diffusion of light from a frosted bulb rather than a clear bulb. Although discrete reflections in the form of echoes are usually unwanted, it may not be desirable to eliminate that sound energy in a room. For example, diffusion can be useful in an auditorium or concert facility to spread sound evenly throughout an audience and ensure that all audience members hear the same sound quality. This min-

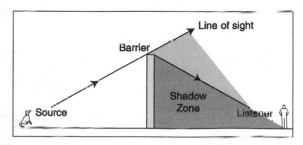

Figure 1.8 Diffraction of sound over a barrier.

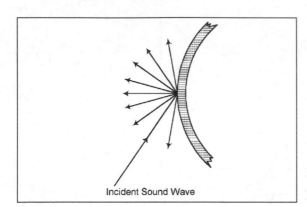

Incident Sound Wave

Figure 1.9 Diffusion of sound.

imizes the potential for bad seats, at least from an acoustical standpoint.

The Decibel

One of the most misunderstood aspects of acoustics is the description of sound in terms of *decibels* (usually denoted as dB). To understand decibels, there are three basic points that must be established:

1. A decibel is defined in terms of a logarithmic ratio. Logarithms are exponents of 10. Therefore, the logarithm of 10 is 1 (since 10 is 10 to the first power), the logarithm of 100 is 2 (since 100 is 10 to the second power), and the logarithm of 1000 is 3 (since 1000 is 10 to the third power). Also, the logarithm of 1 is 0 (since 1 is 10 to the 0 power). The mathematical rules of logarithms are not the same as those for the other systems we normally deal with. A doubling of power on the logarithmic scale translates to adding 3 dB. For example, a 63-dB source has double the power of a 60-dB source, all other conditions remaining unchanged. Thus 60 dB plus 60 dB equals 63 dB, not 120 dB. Along the same lines, a 66-dB source has four times the power of a 60-dB source.

2. The ratio in the decibel's definition has a value that refers to a type of unit, such as power or pressure or intensity. This unit must be specified for a decibel value to have meaning. Using the designation of "level" with one of these qualifiers is one way to satisfy this specification (e.g., sound pressure level identifies pressure as the reference unit). Another way to qualify a decibel value is to state what the reference value specifically is (e.g., "dB re 2×10^{-5} N/m^2" refers

to decibels with reference to 2×10^{-5} newtons per square meter, the standard pressure reference for sound pressure level). Many specifications reference "dB" without any further references. These types of denotations have no meaning. Some examples of common acceptable decibel designations are as follows:

Sound pressure level—SPL, L_p, dB re 2×10^{-5} N/m², dB re 0.0002 µbar

Sound power level—L_W, dB re 1×10^{-12} W

Sound intensity level—$I_{\iota I}$, dB re 1×10^{-12} W/m²

3. The measurement location of the decibel level must be stated. Since sound pressure changes with distance relative to a source, any decibel designation in terms of sound pressure level must also specify a location with respect to the source. To complicate the matter slightly, power is a value that is characteristic of a source and therefore is independent of location. A sound power level, then, should not have a location designation. However, most decibel designations are in terms of sound pressure level. Sound intensity level, though dependent on location, is normally not used for specifications.

The decibel is named for the American inventor Alexander Graham Bell, most noted for his invention of the telephone. The prefix *deci* stands for one-tenth, since units of bels are too coarse to adequately describe our perception of sound. One bel is 10 decibels. The definition of a decibel in the field of acoustics is 10 multiplied by the logarithmic ratio of a source's power to a reference power. Power is proportional to squared sound pressure, and sound pressure level, the quantity most often measured and referenced in architectural acoustics, is then 10 multiplied by the logarithmic ratio of a source's squared pressure (at a specific location) to a squared reference pressure. This reference pressure is approximately the threshold of hearing at 1000 Hz; that is to say, a young person with normal hearing can just hear a 1000-Hz sound above this threshold and cannot hear it below. Hearing sensitivity normally decreases with age and with exposure to high noise levels, in which cases the individual's hearing threshold rises above this value.

When a source's pressure at a certain location is the same as the threshold of hearing, the pressure ratio in the decibel equation becomes 1 and (as already explained) the logarithm of 1 is 0. Therefore, the threshold of hearing has been established as 0 dB. Any negative sound pressure level value is associated with a pressure but has no meaning to our ears since we cannot hear it.

Earlier in this chapter, point and line sources were mentioned. In general, with no obstructions in a sound wave's travel path, the sound pressure level of a sound wave becomes reduced by a factor of 6 dB per doubling of distance from a point source, and by a factor of 3 dB per doubling of distance from a line source. This happens regardless of the sound wave's interaction with any material or medium conditions. The reason for this drop-off rate is that the sound energy (which does not change) expands over an increasing area as the sphere (for a point source) or cylinder (for a line source) of coverage expands with distance from a source (see Figures 1.10 and 1.11). When sound travels outdoors over distances greater than 200 to 300 ft (roughly 50 to 100 m) from a source, other factors influence the drop-off of sound with distance. These include atmospheric absorption, refraction, wind currents, and the effects of terrain. These factors can change sound levels by as much as ±20 dB at locations more than 1000 ft from a sound source.

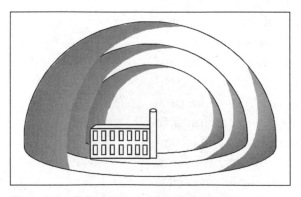

Figure 1.10   Point source spherical wave propagation.

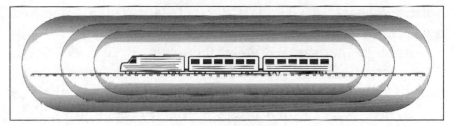

Figure 1.11 Line source cylindrical wave propagation.

Weighting

Our hearing sensitivities are not the same for all frequencies in our hearing range. We are most sensitive to sounds in the 500- to 4000-Hz range. We amplify sounds in the 2000- to 4000-Hz range because of a resonance condition set up by the size and shape of our ear canals. Figure 1.12 shows how our sensitivity to frequency drops off below 500 Hz and above 4000 Hz. At normal conversational levels, a tone at 100 Hz would have to be roughly 20 dB higher in sound pressure level than a tone at 1000 Hz to sound just as loud. Weighting scales were established by the American National Standards Institute (ANSI) to allow us to describe sound (as heard by humans) in terms of a single decibel value covering the entire audible frequency range. These scales are integrated into electronic networks in sound level meters. The most common of these weighting networks is the A-weighted scale. This adjusts frequency levels in accordance with our reactions to sound pressure levels below 70 dB. Although our sensitivities actually change for louder sound pressure levels, the A-weighted scale is used by the Occupational Safety and Health Administration (OSHA) for limits of high-level

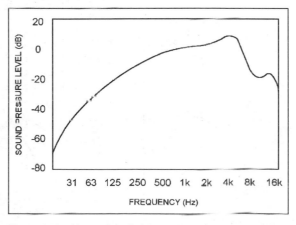

Figure 1.12 Human hearing frequency response.

noise exposure. A-weighted decibels are usually denoted by dBA.

Although there is a B-weighted scale, it is rarely used. dBB values correspond to our hearing mechanism's reaction to sound pressure levels between 70 and 90 dB. The C-weighted scale, which offers minimal reduction in high- and low-frequency ranges, corresponds to our reactions to sound pressure levels above 90 dB. Figure 1.13 shows the characteristics of the A-, B-, and C-weighting scales. As you can see from this figure, our sensitivities to different frequencies level off as sound pressure level increases.

Since the A-weighted scale is the most common one in use, it would be instructive to mention some key points about dBA. Table 1.2 offers common analogies for dBA levels between the human threshold of hearing and the threshold of pain.

The threshold of hearing was already discussed as the minimum sound pressure level that can be heard by the average healthy person. The threshold of pain is the sound pressure level at which most people begin to feel pain from an exposure. Sound pressure levels in this range can cause

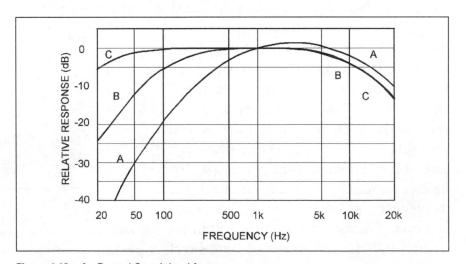

Figure 1.13 A-, B-, and C-weighted frequency responses.

Table 1.2 Common Analogies for dBA Sound Pressure Levels

Common Sound Environment	Sound Pressure Level (dBA)
Threshold of pain	120
300 ft (90 m) from airport runway	110
Typical nightclub	100
OSHA limit for 8-hr exposure	90
Construction area	80
50 ft (15 m) from major highway	70
Urban area during day	60
Quiet suburban area during day	50
Quiet suburban area at night	40
Quiet rural area at night	30
Inside broadcast studio	20
Inside audiometric booth	10
Threshold of hearing	0

immediate permanent hearing loss. Sound pressure levels exceeding 140 dBA are of even greater potential hazard because this is when *acoustic trauma* can occur. Acoustic trauma is the process through which the inner ear hair cells and their supporting structures are physically torn apart. This will obviously cause permanent hearing loss.

General rules of hearing perception are that most people can just notice a change in sound pressure level of 2 to 3 dBA, a 5-dBA change is clearly noticeable, and a 10-dBA change translates to a perceived doubling or halving of loudness. Therefore, assuming the same sound source and listener location, 70 dBA would sound twice as loud as 60 dBA and 80 dBA would sound four times as loud as 60 dBA.

Frequency Bands

Since materials react in different ways to sounds of different frequencies, it is often necessary to have more specific acoustic information as a function of frequency. The most common method of doing this is by designating sound pressure levels in specified frequency bands. These bands are typically represented by a single frequency designation. The most common classifications of these frequency bands are known as *octave bands* and have been defined by ANSI in terms of the center frequencies of 31.5, 63, 125, 250, 500, 1000, 2000, 4000, 8000, and 16,000 Hz. Note that each successive octave band frequency designation is twice the preceding one. The most common octave bands used for measurements and reporting of data are between 125 and 4000 Hz. The human speech frequency range is 500 to 4000 Hz, with vowel sounds toward the lower end of that scale and consonant sounds toward the upper end of that scale.

Now that the basics are covered, we can talk more specifically about controlling sound, both indoors and outdoors.

Sound Control

When we talk about controlling sound, it is often assumed that we are referring to the reduction of sound. However, there are cases where we want to preserve the sound energy but we would like to control its spatial spreading characteristics. The primary ways to reduce sound are through absorption and insulation. Using absorption on an auditorium's side walls may eliminate unwanted reflections but may also eliminate the possibility of some people hearing sound coming from the stage. We therefore must clarify how we plan to control the sound. (For an interesting discussion on good acoustics, see "On the 'Goodness' of Acoustics" in the Technical Addendum.) Redirection and diffusion can have favorable acoustic results for even sound distribution in large rooms.

On the other hand, discussions about noise control usually refer to the reduction of sound (since noise, by definition, is unwanted sound). Figure 2.1 shows what happens to a sound wave that interacts with a room's surface. Part of it is absorbed, part is redirected, and the rest is transmitted through the surface. Each of these phenomena will be

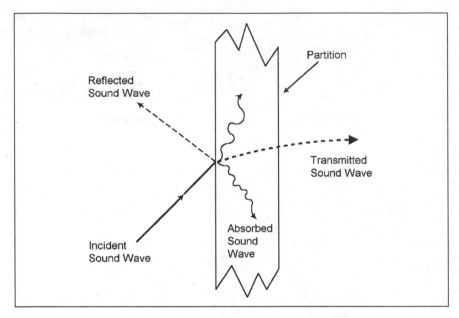

Figure 2.1　Reaction of a sound wave to interaction with a partition.

described here, followed by a discussion of sound reduction methods that utilize these principles.

Absorption

Absorption converts sound energy into heat energy and is used to reduce sound levels *within* rooms. It is not effective in reducing sound *between* rooms. Each material with which a sound wave interacts absorbs some sound. The most common measurement of that is the *absorption coefficient*. The absorption coefficient is a ratio of absorbed to incident sound energy. If a material does not absorb any sound incident upon it, its absorption coefficient is 0. In other words, a material with an absorption coefficient of 0 reflects all sound incident upon it. In practice, all materials absorb some sound, so this is a theoretical limit. If a material absorbs all sound incident upon it, its absorption coefficient is 1. As with the lower limit for absorption coefficients, all materials reflect some sound, so this is also a theoretical limit. Therefore, the limits of absorption coefficients are 0 and 1.

Absorption coefficients vary with frequency. Typical materials used for absorption have absorption coefficients that increase with frequency. They therefore have limited effectiveness for lower frequencies, especially below 250 Hz. There are absorbers that have been designed to absorb these lower frequencies, and these will be discussed shortly. However, for typical cases, it is convenient to use a single number (incorporating multiple-frequency components) to describe the absorption characteristics of a material. This value has been defined by the American Society for Testing and Materials (ASTM) in Standard C423 as the Noise Reduction Coefficient (NRC). The NRC is the arithmetic (as opposed to the logarithmic) average of a material's absorption coefficients at 250, 500, 1000, and 2000 Hz, rounded to the closest 0.05.

Table 2.1 lists the absorption coefficients and NRC values for common materials. Note that the values listed in Table 2.1 are for general reference purposes only, and specific values should be based on manufacturer's specifications. Also note that absorption coefficients and NRC values have no units associated with them. In general, materials with NRC values below 0.20 are considered to be reflective, while those with NRC values above 0.40 are considered to be absorptive. When significant sound energy must be absorbed, as may be the case for eliminating echoes or standing waves, materials having higher absorption coefficients are usually recommended.

A few cautionary notes need to be added. NRC values are convenient to use for rating the absorption characteristics of a material. However, they should be used only when the sound sources of interest are within the 250- to 2000-Hz range. For

Table 2.1 Absorption Coefficients (α_{freq})* and NRC Values for Common Materials

Material	α_{125}	α_{250}	α_{500}	α_{1000}	α_{2000}	α_{4000}	NRC
Painted drywall	0.10	0.08	0.05	0.03	0.03	0.03	0.05
Plaster	0.02	0.03	0.04	0.05	0.04	0.03	0.05
Smooth concrete	0.10	0.05	0.06	0.07	0.09	0.08	0.05
Coarse concrete	0.36	0.44	0.31	0.29	0.39	0.25	0.35
Smooth brick	0.03	0.03	0.03	0.04	0.05	0.07	0.05
Glass	0.05	0.03	0.02	0.02	0.03	0.02	0.05
Plywood	0.58	0.22	0.07	0.04	0.03	0.07	0.10
Metal blinds	0.06	0.05	0.07	0.15	0.13	0.17	0.10
Thick panel	0.25	0.47	0.71	0.79	0.81	0.78	0.70
Light drapery	0.03	0.04	0.11	0.17	0.24	0.35	0.15
Heavy drapery	0.14	0.35	0.55	0.72	0.70	0.65	0.60
Helmholtz resonator	0.20	0.95	0.85	0.49	0.53	0.50	0.70
Ceramic tile	0.01	0.01	0.01	0.01	0.02	0.02	0.00
Linoleum	0.02	0.03	0.03	0.03	0.03	0.02	0.05
Carpet	0.05	0.05	0.10	0.20	0.30	0.40	0.15
Carpet on concrete	0.05	0.10	0.15	0.30	0.50	0.55	0.25
Carpet on rubber	0.05	0.15	0.13	0.40	0.50	0.60	0.30
Upholstered seats	0.19	0.37	0.56	0.67	0.61	0.59	0.55
Occupied seats	0.39	0.57	0.80	0.94	0.92	0.87	0.80
Water surface	0.01	0.01	0.01	0.01	0.02	0.03	0.00
Soil	0.15	0.25	0.40	0.55	0.60	0.60	0.45
Grass	0.11	0.26	0.60	0.69	0.92	0.99	0.60
Cellulose spray (1 in)	0.08	0.29	0.75	0.98	0.93	0.76	0.75

*α_{freq} is the absorption coefficient for a specific frequency (e.g., α_{125} is at 125 Hz).
SOURCES: Acentech data and R. A. Hedeen, *Compendium of Materials for Noise Control* (Cincinnati: National Institute for Occupational Safety and Health, 1980).

sources outside of this range, and especially below this range, materials effective for the specific frequency of interest must be used. Also note that many manufacturers specify NRC and absorption coefficient values that are greater than 1.0. Methods used to measure absorption coefficients can artificially raise their values above 1.0; yet such values inaccurately imply that more energy is absorbed by a material than is incident upon it, which is a physical impossibility. Therefore, any published absorption coefficient or NRC values greater than 1.0 should not be considered as greater than 1.0.

As Table 2.1 shows, the absorption coefficients of most materials increase with frequency. This means that they are not as effective at low frequencies as at higher ones. If absorption is required for frequencies below 250 Hz, special materials must be used. Each of these materials has an air space behind its light or open surface to provide the extra absorption. Two common materials used for this purpose are *Helmholtz resonators* and *diaphragmatic absorbers.* Helmholtz resonators, named for the nineteenth-century German physicist Hermann Ludwig Ferdinand von Helmholtz, are shaped like beverage bottles. They have narrow necks that open to the outside on one end and into a larger air cavity on the other, as is shown in Figure 2.2. As viewed by an incoming sound wave, the air in the narrow neck functions as a mass and the air in the larger cavity functions as a spring. A mass on a spring will resonate at a frequency appropriate to that mass and spring stiffness. Thus, at and near the resonant frequency of such an acoustic chamber, sound will be absorbed from an incoming sound wave, as is shown in Figure 2.3. Such devices incorporated in wall con-

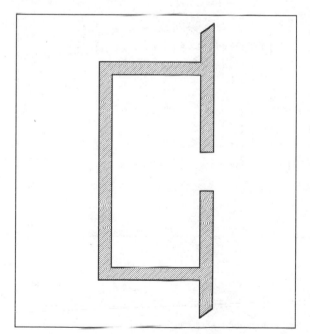

Figure 2.2   General cross section of a Helmholtz resonator.

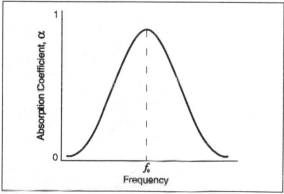

Figure 2.3 General absorption characteristics of a Helmholtz resonator (f_0 is the resonant frequency, which depends on the size of the neck and volume of the cavity).

structions are usually resonant below 250 Hz, depending on the dimensions of the neck and size of the cavity. There are commercially available products that incorporate this design into concrete masonry units. When viewing these products installed as partitions, the wall surfaces have slots in them, as are shown in Figure 2.4. Table 2.1 has a listing for these products which shows their superior absorption in a narrow low-frequency range. The porous and coarse nature of the surface provides modest absorption at higher frequencies also.

Diaphragmatic absorbers work according to principles similar to those for Helmholtz resonators, except they consist of an air space behind a light wall, as shown in Figure 2.5.

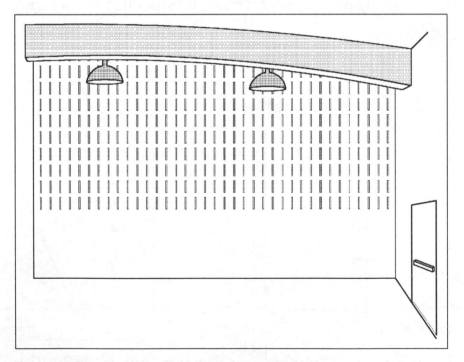

Figure 2.4   Illustration of a wall with Helmholtz resonator blocks.

More of a discussion on the practical nature of this topic is included in "Sound Performance in Public Spaces" in the Technical Addendum.

Reverberation

Absorption is useful in reducing or eliminating unwanted reflections off surfaces. The standing waves discussed in Chapter 1 can be eliminated by covering one of the parallel surfaces with absorptive material. Absorption can also be used constructively to eliminate echoes. The rear walls of auditoriums are prime candidates for absorptive materials since rear walls have the greatest potential to cause echoes. The most common use of absorption, however, is to control reverberation.

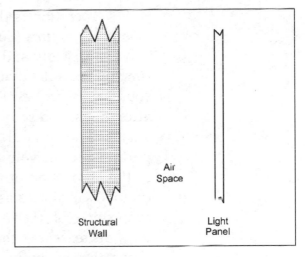

Figure 2.5 Cross section of a diaphragmatic absorber.

Reverberation is the buildup of sound within a room, resulting from repeated sound wave reflections off all of its surfaces. Reverberation can increase sound levels within a room by up to 15 dBA, as well as distort speech intelligibility. Reverberation is desirable for rooms in which music is being played, especially classical and cathedral-style music, to add a pleasant elongation of tones. Therefore, there are different reverberation characteristics that would be appropriate for different room uses.

A discussion of reverberation analysis would not be complete without mentioning the contributions of Wallace Clement Sabine in the late 1800s and early 1900s. Sabine, through his experiments with seat cushions in different facilities, developed the first mathematical definition for describing reverberation characteristics and how to control them. It is from Sabine's efforts that the science of architectural acoustics first developed.

Reverberation is described by a parameter known as the *reverberation time* (denoted RT_{60}). RT_{60} can be defined in two ways: physically and mathematically. Physically, RT_{60} is the time (in seconds) that it takes for a sound source to reduce in sound pressure level (within a room) by a factor of 60 dB after that sound source has been silenced. Mathematically, in what is known as the *Sabine equation*, RT_{60} is directly proportional to the volume of a room and inversely proportional to the absorption of the materials in the room.* This means that RT_{60} increases as the size of a room increases and as the absorption of the room's surfaces decreases. Conversely, RT_{60} decreases as the size of a room decreases and as the absorption of the room's surfaces increases. There are then two principal ways to control RT_{60}: (1) through changing a room's size and (2) through changing the amount of absorption on its surfaces. Although it is possible to vary a large room volume by reducing the volume using partitions, it is more practical to adjust RT_{60} by adding sound absorption using curtains that can be pulled out of sight when not needed or by adding absorptive materials.

Since absorption changes with frequency, so does RT_{60}. Table 2.2 offers generally accepted ranges of RT_{60} for different uses in the midfrequency (500- to 1000-Hz) range. As you can see from the table, lower RT_{60} values are desirable for rooms used mainly for human speech and higher RT_{60} values are desirable for rooms used mainly for music. The optimum midfrequency RT_{60} for a fully occupied room is different for various types of music. Because contemporary orchestral repertoires emphasize late classical and romantic music, the optimum midfrequency RT_{60} for a fully occupied concert hall is usually stated as 1.8 to 2.0 sec.

*Note that the absorption of materials is mathematically defined as the product of the absorption coefficient and the surface area of the material. The units of absorption in English units are called *sabins* (rather than square feet) in honor of Wallace Sabine.

Table 2.2 Optimum Midfrequency RT_{60} Values for Various Occupied Facilities

Type of Facility	Optimum Midfrequency RT_{60} (sec)
Broadcast studio	0.5
Classroom	1.0
Lecture/conference room	1.0
Movie/drama theater	1.0
Multipurpose auditorium	1.3–1.5
Contemporary church	1.4–1.6
Rock concert hall	1.5
Opera house	1.4–1.6
Symphony hall	1.8–2.0
Cathedral	3.0 or more

Optimum RT_{60} values generally increase by 10 percent of the values in Table 2.2 for each halving of frequency below 500 Hz and decrease by 10 percent of the values in Table 2.2 for each doubling of frequency above 1000 Hz. A room with a low (less than 0.8 sec) RT_{60} is called a "dead" room and a room with a high (greater than 1.7 sec) RT_{60} is called a "live" room. Multipurpose facilities should have RT_{60} values between the live and dead range limits.

Two interesting case studies involving reverberation control in churches are included in later sections—one on the Princeton University Chapel in Chapter 3 and one in the article entitled "Gothic Sound for the Neo-Gothic Chapel of Duke University" in the Technical Addendum. Each case deals with the same issue of the chapel's not having enough reverberation for the desired environment.

Redirection

Although absorption is necessary in many circumstances, eliminating reflections is not always a useful thing to do. This is of key importance in rooms where an audience is listening to a performance or lecture. In this type of room, it is desirable that all audience members hear the sound not only clearly, but without preference to seating location. Without an electronic sound system, this can be accomplished only by reflections off side walls and ceilings. Sharp echoes can be eliminated by avoiding smooth, flat, reflective surfaces and having irregular and convex surfaces to diffuse the sound evenly throughout the audience. For smaller rooms such as recording studios that require diffusion, special commercial sound-diffusing panels called QRDs (standing for *quadratic residue diffusers*) are available. However, larger spaces do not require the intricate designs or extensive budgets to accomplish the same type of acoustic environment.

Concave surface shapes should be avoided. Concave reflective surfaces focus sound in certain areas and defocus sound from others, causing hot spots where sound is concentrated and dead spots where sound cannot be heard. Concave reflective surfaces should be avoided for this reason. If aesthetics dictate the need for a concave reflective surface, it would be best to install an absorptive or diffusive surface (as needed) and cover it with acoustically transparent material in the concave shape.

Reflective rear walls in auditoriums are notorious for generating echoes because of their associated large sound travel path differences. For this reason, reflective surfaces should be avoided for rear walls. Reflective surfaces are beneficial, especially for concert halls, when they are close to the stage and along side walls. Reflective surfaces close to the stage assist in several ways, by sending sound out into the audience rather than allowing it to be lost behind the stage and by enhancing the sound through lateral reflections off side walls to spread the sound more evenly throughout an audience. Another benefit of reflective surfaces near the stage is that they allow the performers to hear each other, something that is critical to concert performances. These so-called *early reflections* are usually generated by shells on the stage or by hanging reflective panels, called *clouds*. Examples of these are provided in later chapters.

Insulation

The description of the insulation of sound is similar in many ways to the description of the absorption of sound. As for absorption, there is a *transmission coefficient* that ranges from the ideal limits of 0 to 1. The transmission coefficient is the unitless ratio of transmitted to incident sound energy. Unlike the absorption coefficient, however, the limit of 1 is practically possible since a transmission coefficient of 1 implies that all of the sound energy is transmitted through a partition. This would be the case for an open window or door, where the sound energy has no obstruction to its path. The other extreme of 0 (implying no sound transmission), however, is not a practical value since some sound will always travel through a partition.

Unlike absorption, the principal descriptor for sound insulation is a decibel level based on the transmission coefficient. This value is known as the *transmission loss* (TL) and is based on the logarithm of the mathematical reciprocal of (1 divided by) the transmission coefficient. The transmission loss can be loosely defined as the amount of sound reduced by a partition between a sound source and a listener. The complete sound reduction of a partition between two rooms also takes into account the partition's installation and the absorptive characteristics of the rooms. However, TL is the quantity that is typically reported in manufacturers' literature since it is measured in a laboratory independent of the installation.

Since the logarithm of 1 is 0, the condition in which the transmission coefficient is 1 translates to a TL of 0. This concurs with the notion that an open air space in a wall allows the free passage of sound. The practical upper limit of TL is roughly 70 dB.

As for absorption, TL is frequency dependent. Typical partitions have TL values that increase with increasing frequency. There is also a single-number rating for TL that takes the entire frequency spectrum into account, established by ASTM Standard E413. This value, known as the Sound Transmission Class (STC), is not derived by the simple averaging method used for NRC values. Instead, the TL frequency spectrum (plot of TL versus frequency) is matched to a standard curve within the limits imposed by the ASTM standard. Note that STC values have no units associated with them, but they are based on decibels. Similar to the NRC, STC is useful to describe the sound insulation efficiency of a partition over the human speech frequency range of roughly 500 to 2000 Hz. If sound insulation outside of that frequency range (especially for frequencies below 250 Hz) is required, the TL values relevant to the frequency range of interest must be used.

Table 2.3 offers TL and STC values for common partitions. As with Table 2.1, these values are for general reference purposes only, and specific values should be based on manufacturer's specifications. Table 2.4 gives some meaning to the numbers by listing sound privacy ratings for different ranges of STC values.

Up to this point we have been discussing homogeneous partitions. Actual designs, however, include walls composed of different materials, such as those with windows and doors. Each of these wall components has different associated TL characteristics. Wall components having lower TL characteristics than the rest of the wall can significantly degrade a wall's sound reduction effectiveness. For example, placing a 4-ft^2 window with a rating of STC 25 in a 100-ft^2 wall of STC 55 reduces the composite STC to a value of 39.

Air gaps (for example, around doors) are notorious for compromising the sound reduction effectiveness of walls. Table 2.5 shows the TL degradation of varying sizes of air gaps in a wall originally rated at a TL of 45 dB. As is shown

Table 2.3 Transmission Loss (TL$_{freq}$)* and STC Values for Common Partitions

Partition	TL$_{125}$	TL$_{250}$	TL$_{500}$	TL$_{1000}$	TL$_{2000}$	TL$_{4000}$	STC
½-in drywall on wooden studs	17	31	33	40	38	36	33
½-in drywall on wooden studs with 2 in of insulation	15	30	34	44	46	41	37
Double layer of ½-in drywall on wooden studs	25	34	41	51	48	50	41
½-in drywall on staggered wooden studs	23	28	39	46	54	44	39
½-in drywall on staggered wooden studs with 2 in of insulation	29	38	45	52	58	50	48
½-in drywall on metal studs	22	27	43	47	37	46	39
½-in drywall on metal studs with 2 in of insulation	26	41	52	54	45	51	45
Concrete masonry units	34	40	44	49	59	64	49
Open-plan office partition	10	12	12	12	12	11	12
Brick wall	32	34	40	47	55	61	45
½-in drywall inside/1-in stucco outside on wooden studs	21	33	41	46	47	51	42
Single-paned ⅛-in glass	18	21	26	31	33	22	26
½-in laminated glass	31	34	38	40	37	46	40
Double-paned ⅛-in glass with 2-in air gap	13	25	35	44	49	43	37
Hollow wooden door	14	19	23	18	17	21	19
Solid wooden door	29	31	31	31	39	43	34
Hollow metal door	24	23	29	31	24	40	28
Filled metal door	26	34	40	48	44	52	43
Wood joist floor/ceiling	23	32	36	45	49	56	37
Concrete slab floor	32	38	47	52	57	63	50
Floating floor	30	44	52	55	60	65	55
Wood plank shingled roof	29	33	37	44	55	63	43
Wood plank shingled roof with ½-in drywall ceiling, 4 in of insulation	35	42	49	62	67	79	53
Corrugated steel roof with 1 in of sprayed cellulose	17	22	26	30	35	41	30

*TL$_{freq}$ is the transmission loss for a specific frequency (e.g., TL$_{125}$ is at 125 Hz).

SOURCES: Acentech data and R. A. Hedeen, *Compendium of Materials for Noise Control.* (Cincinnati: National Institute for Occupational Safety and Health, 1980).

Table 2.4 Sound Privacy Associated with STC Ratings

STC Range	Sound Privacy
0–20	No privacy (voices heard clearly between rooms)
20–40	Some privacy (voices heard in low background noise)
40–55	Adequate privacy (only raised voices heard in low background noise)
55–65	Complete privacy (only high-level noise heard in low background noise)
70	Practical limit

in the table, an air gap just one-tenth of one percent the size of a wall can lower the TL rating from 45 to 30 dB. This demonstrates the importance of sealing walls, avoiding air gaps, and placing airtight seals around doors.

A rule that governs most of the TL spectrum of homogeneous partitions, known as the *mass law,* states that TL increases at a rate of 6 dB with each doubling of mass and with each doubling of frequency. This rule can cause it to be impractical to solve sound privacy issues with homogeneous partitions. For example, if a 1-ft-thick concrete wall does not provide enough sound insulation for a specific situation, doubling the thickness of that wall to 2 ft would only offer an additional 6 dB of sound reduction. The most practical way of avoiding this kind of issue is by using multilayered partitions.

Multilayered partitions comprise layers of different materials. Each time sound passes through a different material, it is reduced. Therefore, this method can be used to reduce costs and space restrictions while providing adequate sound reduction. A sharp change in density of material is most effective in this manner. Air spaces between wall sections and materials are effective by setting up such environments

Table 2.5   Transmission Loss Reduction as a Function of Air Opening*

% of Wall Area Having Air Opening	Resultant Wall TL (dB)	Resultant Reduction in TL (dB)
0.01	39	6
0.1	30	15
0.5	23	22
1	20	25
5	13	32
10	10	35
20	7	38
50	3	42
75	1	44
100	0	45

*Based on original wall TL of 45 dB.

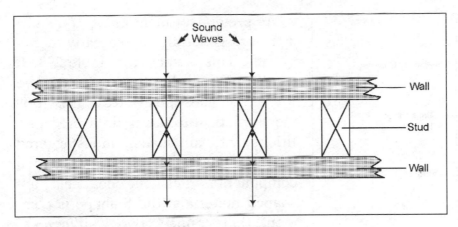

Figure 2.6 Sound waves transmitted through studs in a wall.

and also by breaking any rigid connections between sides of a partition. A rigid connection can provide a vibration channel for sound to pass through with little reduction. For example, the noise reduction effectiveness of a studded wall filled with fiber insulation between studs can be short-circuited because sound will travel through the studs to the other side of the wall, as shown in Figure 2.6. Staggered studs for the same wall can provide significantly higher TL while sacrificing minimal space (see Figure 2.7).

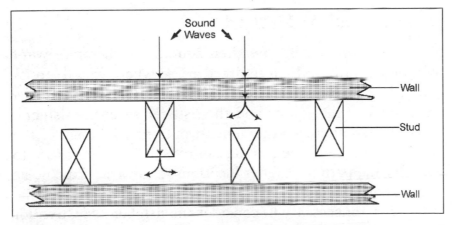

Figure 2.7 Sound transmission being dissipated through staggered studs.

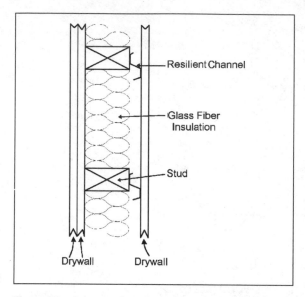

Figure 2.8   Cross section of a multilayered partition.

As previously mentioned, transmission loss generally increases with frequency. This is shown in Table 2.3. If significant sound reduction is required for frequencies below 250 Hz, it is most practical to use multilayered partitions with air spaces and staggered studs or resilient channels between components. Resilient channels are shaped materials (the S shape is common) that connect two wall components. Only one side of their shape touches each side of the wall, and they effectively reduce vibrational energy from traveling through them. Figure 2.8 shows a cross section of a multilayered partition with a resilient channel. Another option for greater transmission loss at low frequencies is massive concrete slabs. For further discussion on these topics, see "Blocking the Bombardment of Noise" in the Technical Addendum.

With these basic concepts explained, we are now in the position to discuss common noise reduction methods.

Noise Reduction Methods

Noise by definition is unwanted sound. We therefore want to eliminate, rather than redirect, noise when we talk about controlling it. Noise can be controlled at its source, in the path between the source and the listener, or at the listener. Table 2.6 summarizes the general options available. If the noise can be controlled at its source, it is unnecessary to consider the path or listener locations. Likewise, if the noise can be controlled in the path between the source and listener, it is unnecessary to consider the listener's location for noise control measures.

Table 2.6 Noise Reduction Options

Control at the Source	Control in the Path	Control at the Listener
Maintenance.	Enclose source.	Relocate listener.
Avoid resonance.	Install barrier.	Enclose listener.
Relocate source.	Install proper muffler.	Have listener use hearing protection.
Remove unnecessary sources.	Install absorptive treatment.	Add masking sound at listener's location.
Use quieter model.	Isolate vibrations.	
Redesign source to be quieter.	Use active noise control.	

The options for noise control at the source are generally self-explanatory. Although they are the preferred noise control options, they are often impractical logistically or economically. Most often, noise control options are limited to the path between the source and the listener and at the listener. Given many misconceptions about these options, it is useful to discuss some of them further.

Enclosures

Enclosures can be effective at reducing noise levels, as long as they are designed properly. Consider the following points when designing noise enclosures (also illustrated in Figure 2.9*a* through *d*):

1. The enclosure must completely surround the noise source, having no air gaps. As is mentioned in the section entitled "Insulation" in this chapter, air gaps can significantly compromise the noise reduction effectiveness of partitions. Think of waterproofing. If water can leak through a partition, so can noise. An enclosure with any side open is not an enclosure but a barrier, and the noise reduction effectiveness of barriers is limited by diffraction to 15 dBA, independent of the barrier material. Enclosures, on the other hand, can provide up to 70 dBA of reduction.

2. The enclosure must be isolated from floors or any structural members of a building. An enclosure covering the sides and top of a noise source but having the

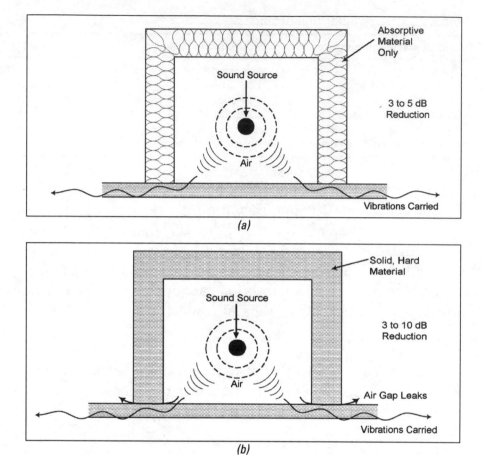

Figure 2.9 Effectiveness of different enclosure designs: (*a*) absorptive material only, placed on solid floor; (*b*) rigid material only, placed on solid floor;

bottom open (since the source is sitting on the floor or ground) can compromise its effectiveness for several reasons. First, the chances of the sides of the enclosure perfectly sealing to the ground are slim, and therefore, air gaps would result. Second, vibrations will be carried along the ground or floor since the source is in direct contact with it. The only way to reduce these vibrations is to vibrationally isolate the source from the ground or floor using tuned springs (appropriate for the source), pads, or the bottom of a multilayered enclosure.

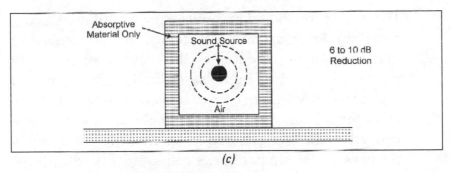

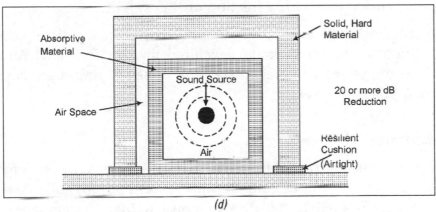

Figure 2.9 (*cont.*) (*c*) absorptive material completely surrounding sound source; (*d*) combination of methods for a multilayered enclosure.

3. The enclosure should not consist of only sound-absorptive material. Sound-absorptive material can be effective in reducing noise when it is used as part of a multilayered enclosure (on the inside); however, absorptive material on its own is not effective in reducing noise. The main purpose of absorptive material is to control reflections within spaces, not to control sound transmission out of spaces.

4. The enclosure must consider that some sources require ventilation. This cannot translate to leaving a simple opening in the enclosure without severely compromising the noise control effectiveness of the enclosure. Ventilation systems must be developed that minimize noise transmission.

5. The enclosure should be built using multilayered construction for maximum efficiency. As is mentioned under "Insulation," doubling the mass of an enclosure would add 6 dB to its noise reduction effectiveness. This can easily lead to excessive weight for an effective homogeneous enclosure. As for single partitions, multilayered enclosures can add more than 20 dB of effectiveness under similar space requirements to massive enclosures with a fraction of the weight.

For more details on these issues in construction, see "Quieting the Multifamily Dwelling" in the Technical Addendum, which explores further the issues of sound privacy on a practical level.

Barriers

A barrier is contrasted from an enclosure by it being open to the air on at least one side. Because of diffraction, noise barriers are limited to 15 dBA of noise reduction capability, independent of the material. This limited effectiveness is compromised even more if there are reflective ceilings above the barrier because sound reflected off the ceiling minimizes the barrier's effectiveness. Therefore, wherever noise barriers are used indoors, an absorptive ceiling should be installed above them. It is also important to have no air spaces within or under the barriers, since this will compromise their already limited effectiveness.

The noise reduction effectiveness of barriers is typically rated by the *insertion loss* (denoted IL). IL is the simple reduction in sound pressure level, at a specific location, with a barrier in place. In other words, IL is the difference between conditions with and without a barrier.

To provide any insertion loss, a barrier must break the line of sight between the sound source and the listener. In other words, if you can see a sound source on the other side of a barrier, that barrier is providing no sound reduction

(from that source) for you. Breaking this line of sight typically provides a minimum of 3 to 5 dBA of insertion loss,
with insertion loss increasing as one goes further into the
shadow zone of the barrier.

Mufflers

Mufflers are devices that are inserted in the path of ductwork or piping with the specific intention of reducing sound
traveling through that conduit. The effectiveness of mufflers
is typically rated using insertion loss. Mufflers must be
designed for each purpose to preserve the required pressure
characteristics. For that reason, each muffler is unique to its
installation. The design of mufflers should therefore be del-
egated to those having experience in that regard.

Absorptive Treatment

Absorptive treatment within a room can reduce reverberation in a room and, in this process, reduce noise levels by up
to 10 dBA. Bear in mind, though, that absorptive treatment
is effective only for reducing reverberation within a room
and not for transmission of sound between rooms. For more
information on the uses of absorption for noise control, see
"Soaking Up Sound: Properties of Materials That Absorb
Sound" in the Technical Addendum. In addition, two discussions on restaurants in later sections—one in a case study in
Chapter 4 (on No. 9 Park) and one in the Technical Addendum article entitled "Quieting the Noisy Restaurant"—
demonstrate the value of absorption in controlling noise.

Vibration Isolation

Mechanical equipment can generate vibrations that can
travel through a building's structural members to affect
remote locations within a building. It is therefore prudent to
isolate any heavy equipment from any structural members
of buildings. This can be accomplished by mounting the
equipment on springs, pads, or inertia blocks; however, the

selection of specific isolating equipment should be performed by a specialist trained in vibration analysis. The main reason for this is that each vibration isolation device is tuned to a specific frequency range. If this is not matched properly with the treated equipment, the devices can amplify the vibrations and cause more of a problem than would have occurred without any attempted treatment. For further discussion of this topic, see "Vibration Control Design of High-Technology Facilities" in the Technical Addendum.

Active Noise Control

Passive noise control involves all of the noise control methods discussed so far in this book, in which the sound field is not directly altered. Active noise control involves electronically altering the character of the sound wave to reduce its level. In this case, a microphone measures the noise and a processor generates a mirror image of (180° out of phase from) that source. This mirror image is then reproduced by a loudspeaker in the path of the original sound. This new sound cancels enough of the original signal to reduce levels by up to 40 dB in the appropriate circumstances. Figure 2.10 illustrates this procedure. Although this is a very powerful noise control tool, active noise control is practical only in local envi-

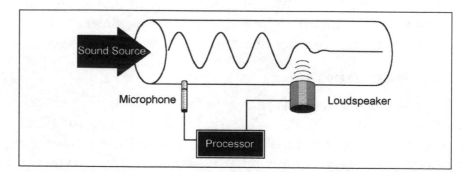

Figure 2.10 The active noise control process.

ronments and for tonal frequencies below 500 Hz. Ventilation ducts are ideal candidates for active noise control systems. This is because they are enclosed environments and because their dominant noise is often low-frequency pure tones (associated with the fan characteristics).

Masking

As long as background sound levels are low in a building, one way of reducing a noise problem is to add a more pleasing sound to the environment that covers up the noise or makes it less noticeable. This is especially desirable in open-plan offices where people can clearly hear activities in other offices and areas (since their offices are not completely enclosed). Any desirable sound can provide masking, but most often electronic masking systems consist of loudspeakers placed between dropped ceilings and structural ceilings. These loudspeakers are connected to signal processors that are set to generate sounds similar to those generated by typical heating, ventilating, and air-conditioning (HVAC) systems. Although many people think of masking system sounds as white noise, which has an equal amount of energy in all audible frequencies, a typical masking system frequency response is more like that shown in Figure 2.11, where less emphasis is placed on higher frequencies. Whatever the response, it is advisable to have an electronic masking system installed by a contractor with experience in this area. If the system is set at too loud a level or with a harsh frequency response, the environment may wind up being more unpleasant with the masking system than without.

The results of a survey are included in "Sound Masking: The Results of a Survey of Facility Managers" in the Technical Addendum, which provides interesting insights into the reactions and plans for sound masking in many types of buildings.

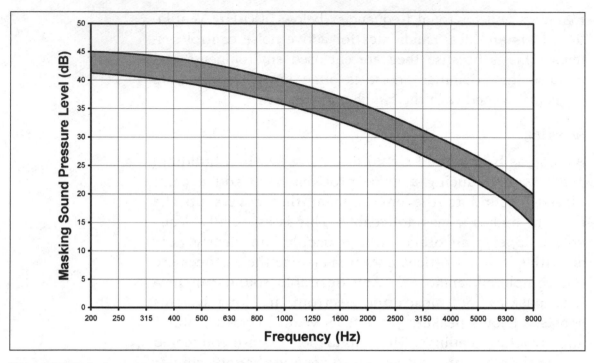

Figure 2.11   Typical sound-masking system levels and frequency response.

Rating Interior Environments

There are two standardized methods for rating the background sound levels inside rooms set by HVAC systems. Each of these methods is referenced in ANSI Standard 12.2 (*American National Standard Criteria for Evaluating Room Noise*, 1995) and the 1999 edition of the *American Society of Heating, Refrigerating and Air-Conditioning Engineers* (*ASHRAE*) *Handbook*. These are the NCB and RC room criterion curves. The NCB curves are updated versions of the widely used NC (noise criterion) curves (the original rating method for interior environments derived by acoustician Leo Beranek in the 1950s), different in that they are extended down to the 31.5- and 63-Hz octave bands. "The Sound of NC-15" in the Technical Addendum goes into more practical detail on this rating.

As for the RC method, the NCB method was updated to better consider low-frequency noise generated by HVAC systems. In each case, standardized curves are used in conjunction with measured noise levels in a room to yield a single-number (NCB or RC) rating for HVAC noise. This single number is then rated against design guidelines established in the two aforementioned references. The RC curves were developed primarily for use in builders' specifications. For these curves, the levels in the octave bands below 500 Hz are much lower than for the NCB curves. In setting these levels, fluctuations or surges at the ventilation outlets are allowed. These fluctuations are not permitted for NCB ratings. A new set of noise criterion curves is currently being considered for standardization, having decibel values below 500 Hz that lie between the RC and NCB curves, depending on the amount of fluctuations or surges in the air supply that are permitted or measured.

Outdoor Sound Control

Outdoor sound control normally deals with noise reduction issues. Noise reduction options at the source and at the listener are the same whether that source is indoors or outdoors (see Table 2.6). The difference in treatments lies in the path between the source and the listener. In outdoor environments, practical noise reduction choices are limited to the options of buffer zones, barriers, and masking. For example, enclosures are often impractical for large or moving outdoor sources (such as traffic on highways or rail lines).

Buffer Zones

Space is one aspect of the outdoor environment that is often more available than in the indoor environment. As was mentioned in Chapter 1, sound generally dissipates at a rate of 3 to 6 dB per doubling of distance from a source within 200 to 300 ft of that source. Its decay rate beyond that is highly variable depending on the atmospheric (mainly temperature variations, wind currents, and humidity) and terrain conditions between the source and listener. However, sound levels generally decrease with increasing distance from a source. Therefore, the greater the distance that can be placed between an objectionable sound source and a listener, the better.

It is best to avoid placing an objectionable sound source near a still body of water that lies between the source and a listener because refraction effects will cause the sound to travel across the body of water with little reduction. Although wind currents are constantly changing, it is best to avoid locating a noise-sensitive building (such as a residence, house of worship, health care facility, or school) in the prevailing downwind direction of a noise source. As with temperature variations, shadow zones are set up upwind of a noise source and sound travels farther outdoors with the wind.

Barriers

Noise barriers can be effective at reducing noise levels within 200 ft of a sound source but not beyond that distance. However, that effectiveness is limited by the phenomenon of diffraction mentioned in Chapter 1. Independent of the material, barriers provide a maximum of 15 dBA of noise reduction to a listener. The most important design aspects of the barrier are that it is solid, it can stand up to the elements, and it breaks the line of sight between the source and the listener. Any air gaps will compromise a barrier's already limited effectiveness. Figure 2.12*a* and *b* show the noise reduction effectiveness of outdoor barriers. This is also relevant to indoor barriers, for open-office or partial enclosure designs, as long as the ceiling is absorptive. Otherwise, reflections off the ceiling will compromise the limited effectiveness available.

It is often thought that trees or other types of vegetation between a source and a listener will provide a barrier effect.

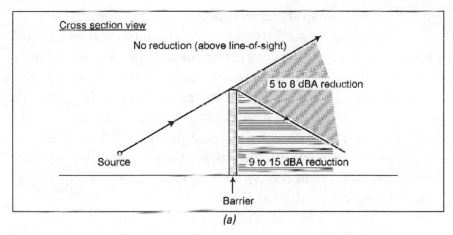

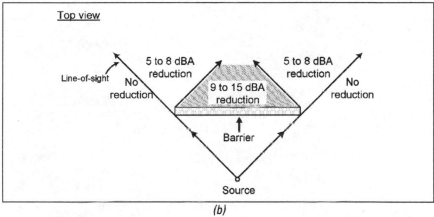

Figure 2.12 Noise reduction effectiveness of barriers, assuming no nearby reflective surfaces: (*a*) cross-sectional view; (*b*) top view.

However, studies confirm that vegetation has minimal effect on reducing noise, unless it is in the form of a dense forested area of evergreens more than 100 ft thick. The only natural design that will serve as an effective noise barrier is a berm or hill.

Masking

Masking systems work best when they blend with the environment to the point at which they go unnoticed. The electronic spectrum referenced in Figure 2.11 is ideal to blend with or supplement typical indoor sounds, such as HVAC noise. Acceptable sounds outdoors include more natural sounds, such as running water or rustling leaves. As is mentioned in the earlier section on noise reduction methods, masking sounds are added to an environment (e.g., small parks or patios) to cover unwanted sounds with acceptable ones. Outdoor fountains are not only effective in masking sound, but add aesthetically to an area. Any other natural masking sounds would have to be added electronically using weatherproofed loudspeakers.

Design Tools and Examples

When designing any facility for acoustics, the first question that must be answered is, "What is its primary purpose?" This purpose will dictate the desired acoustical parameters and treatments for each space. The rest of this book contains case studies emphasizing key acoustic principles, each section being introduced by a list of design considerations for the category of acoustical issue.

CHAPTER 3

Room Acoustics

The key acoustical issues covered in this chapter are reverberation, room shapes, and interior noise control.

Reverberation

Design Tools

General Considerations:

1. Size—minimize the room volume where low reverberation times are necessary (speech auditoriums) and choose the proper larger room volume for cases where medium or high reverberation times are required (halls for music).

2. Absorption—add absorptive materials to reduce reverberation and add reflective or diffusive materials to add reverberation.

3. Low-frequency absorption—use Helmholtz resonators, diaphragmatic absorbers, or plenum absorbers for large rooms having reflective surfaces and in need of speech intelligibility.

4. Speech intelligibility—for large rooms, use a distributed sound system with appropriate delays between loudspeakers and focused low-level loudspeaker systems for reverberant rooms.

5. Sound systems are often used in rooms where contemporary music is being played. While we would like the natural acoustics of the room to provide the necessary sound quality, these sound systems are often set at such high levels that they negate the effect of any architectural designs. Artificial reverberation can also be set in these sound systems. It is therefore advisable to have as much absorption as possible in rooms where contemporary amplified music will be played.

Room Purpose (Full Occupancy)	Design Considerations	Potential Problems	Solutions
Lecture	Minimize RT_{60} with even spreading to audience—minimize room size, maximize absorption.	Uneven spreading, excessive reverberation for large rooms	Use reinforcing sound system; place reflective surfaces close to source to direct sound to audience; add significant absorption to all room surfaces.
Lecture and contemporary music	Keep RT_{60} less than 1.5 sec.	Excessive reverberation for large rooms	Be generous with absorption but allow some reflective surfaces on side walls and ceiling to direct sound to the audience.
Contemporary music	Keep RT_{60} between 1.3 and 1.6 sec.	Reverberation time too short or too long	Allow for absorption on floors, rear walls, and seats.
Contemporary and classical music	Keep RT_{60} between 1.6 and 1.8 sec.	Reverberation time too short or too long	Use moderate amounts of absorption, but leave ceiling and side walls reflective and diffusive.
Classical and romantic music	Keep RT_{60} between 1.8 and 2.0 sec.	Reverberation time too short or too long	Use some absorption on floors and rear wall.
Classical and cathedral music	Keep RT_{60} between 2.0 and 2.2 sec.	Reverberation time appropriate for one but not the other	Confine absorption to seating area and rear wall; have diffusion elsewhere; consider adjustable acoustics.

Room Purpose (Full Occupancy)	Design Considerations	Potential Problems	Solutions
Cathedral music	Keep RT_{60} above 4.0 sec—maximize room size, minimize absorption.	Reverberation time too short for smaller rooms or those with absorptive surfaces	Minimize absorptive surfaces and use uneven and convex wall shapes for diffusion; consider electronic reverberation enhancement systems for smaller spaces.
Lecture and cathedral music	Keep RT_{60} above 4.0 sec with electronic assistance for lecture.	Unintelligible speech	Position many loudspeakers close to listeners at low levels for speech.
Lecture and classical music	Set RT_{60} no higher than 1.8 sec with electronic assistance for lecture.	Unintelligible speech	Position many loudspeakers close to listeners at low levels for speech.

Case Study: School Gymnasium—Pierce School (Brookline, MA)

School gymnasiums are typically large indoor spaces having hard, reflective surfaces. These generate high reverberation times, which make speech intelligibility difficult, if not impossible, in the room. Add to the room an electronic sound system for public address announcements and speech intelligibility can become impossible. The solution to this problem lies in the amount of absorption that can be added. Aesthetic and durability issues often cause restrictions, so they must also be considered.

It is always more practical to design absorption into gymnasiums rather than to attempt to fix the acoustics after the gymnasium has been built. This is because low-frequency absorption is much easier to implement when it is designed into the room, and, since gymnasiums tend to be large rooms, they generate significant low-frequency reverberation.

The Pierce School was opened in 1972, sized for approximately 600 elementary school students. The gymnasium has two components, an upper gym for regular classes and a lower gym for students with special needs. The upper gym is approximately 6000 ft^2 (550 m^2) with a 23-ft (7-m) ceiling; the construction is exposed wood deck on steel joists. The upper walls are concrete block and the lower walls have wood-faced lockers and pullout bleachers.

The midfrequency reverberation time in the gym was roughly 6 sec with one class, which made speech intelligibility difficult. Instructors could not be heard over the noise generated by the students and they could not be understood even if they could be heard. Within one week after the school was opened, renovations and studies had begun. Absorptive treatment was needed to lower the reverberation time in the room. Glass fiber boards covered with 1-in-thick (2.5-cm-thick) shredded wood fiber form boards were chosen to cover about 70 percent of the ceiling, applied in a manner to allow the wood deck to still be recognized. This treatment was also applied as a band, 8 ft (2.4 m) high, around the upper walls of the gym, just below where the steel joists came into the wall.

The results are dramatic, in that speech can be understood in the room. Figures 3.1 and 3.2 show the results of the design. One

Figure 3.1   Pierce School gymnasium, with absorptive panels on the side walls and ceiling. *(Photograph by Joan McQuaid.)*

Figure 3.2 Closer view of the ceiling panels. *(Photograph by Joan McQuaid.)*

note about this installation is that some fiberboards were mounted behind basketball backboards and along the side walls of the stairwell leading out of the gymnasium. As can be seen in Figures 3.3 and 3.4, these types of boards are not appropriate in these areas because they are not durable and basketballs and students can easily destroy them. A more appropriate choice of treatments in these areas would have been glass fiber encased in a perforated metal housing. With the appropriate opening size in the metal, these panels would provide absorption similar to that of the original panels while being much more tolerant of impacts or vandalism.

Figure 3.3   Chipped absorptive panels behind basketball backboard. *(Photograph by Joan McQuaid.)*

Figure 3.4   Stairwell absorptive panels, with a section torn off. *(Photograph by Joan McQuaid.)*

Case Study: Courtroom—Somerset County Courthouse (Somerville, NJ)

The Somerset County Courthouse is the county seat, located in Somerville, New Jersey. It was designed in a neoclassical style, with a domed ceiling, balconies on three sides, and a podium for the presiding judge. Its large volume and curved surfaces created acoustical problems, especially an excessive reverberation time for speech and sound focusing from the concave dome. The management wanted the historic look of the courthouse to be preserved, but the acoustics of the facility had to be improved for speech intelligibility. This translated to adding absorptive materials to the room's surfaces, especially the domed ceiling.

The best way to maximize absorption while preserving the aesthetics was to apply a spray-on mineral-based absorptive material. This was applied to side walls and the ceiling with the result of lowering the reverberation time to an acceptable level, minimizing hot- and dead-spot focusing from the dome ceiling, and providing a look that was as smooth as the original plaster from a distance. This is shown in Figures 3.5 through 3.9.

Figure 3.5 Somerset County Courthouse, view from the rear of the main courtroom toward the front. *(Photograph by Sharon Paul Carpenter.)*

Figure 3.6   View from the front toward the rear. *(Photograph by Sharon Paul Carpenter.)*

Figure 3.7   Side walls of the courtroom. *(Photograph by Sharon Paul Carpenter.)*

Figure 3.8 Dome ceiling of the courtroom. *(Photograph by Sharon Paul Carpenter.)*

Figure 3.9 Close view of the walls, showing the absorptive spray-on material. *(Photograph by Sharon Paul Carpenter.)*

Case Study: Cathedral—Princeton Chapel (Princeton, NJ)

The Princeton University Chapel was constructed in the late 1920s in a collegiate Gothic style popularized by Cram and Fergusson. Since the architects knew of the high reverberation times that would be generated by such a large space (and their effect on speech intelligibility within the space), they made generous use of an artificial stone that was quite porous and sound absorptive. The result was a midfrequency reverberation time of roughly 3.5 sec, which was marginally acceptable for speech intelligibility but not appropriate for the organ music. This is then a case of too low of a reverberation time. To increase the reverberation time, it was decided to paint the porous surfaces to seal them and provide more of a reflective surface. This was not performed until the mid-1980s. The result was an increase in midfrequency reverberation time to more than 5 sec, improving the quality of the organ

Figure 3.10   Princeton Chapel, view from the rear to the front. *(Photograph by James Cowan.)*

music dramatically within the space. The same kind of problem and solution were encountered for a similar building on the campus of Duke University, where midfrequency reverberation times were increased to roughly 7 sec. This case is discussed in more detail in the Technical Addendum article entitled "Gothic Sound for the Neo-Gothic Chapel of Duke University."

This improvement for the organ music was a detriment for speech intelligibility. To solve this problem, loudspeakers were installed in the backs of pews toward the front of the chapel. Column loudspeakers were mounted along the sides in the nave, each set at a low enough level to be heard yet not to generate unnecessary reverberation. Because of the spacing and large number of loudspeakers, they had to be controlled with delaying circuits to allow the sound to arrive at audience members' ears without the distortion of arrival delays from other loudspeakers. Figures 3.10 through 3.14 show the inside of the chapel and its most striking features.

Figure 3.11 Princeton Chapel, view from the front to the rear. *(Photograph by James Cowan.)*

Figure 3.12 Chapel organ and front walls. *(Photograph by James Cowan.)*

Figure 3.13 Architectural details of side walls. *(Photograph by James Cowan.)*

Figure 3.14 Loudspeakers in the backs of pews to enhance speech intelligibility in the front seating sections of the chapel. *(Photograph by James Cowan.)*

Room Shapes

Design Tools

Wall Shape	Design Considerations	Potential Problems	Solutions
Flat	Reflective surfaces can cause acoustic anomalies.	Echoes, uneven sound spreading	Angle walls to direct sound to audience; use absorptive or diffusive materials on walls that would cause 100-ft path delay.
Parallel flat walls	Reflective surfaces can cause acoustic anomalies.	Echoes, flutter echoes, standing waves	Avoid parallel reflective walls or treat one wall with absorptive material.
Concave	Minimize reflective domes and other concave reflective surfaces.	Hot and dead spots	Either eliminate dome and concave surfaces, use absorptive spray-on material, or cover absorptive material with concave, acoustically transparent material.
Convex or uneven	Allow for even spreading of sound (diffusion).	Excessive reverberation	Add absorption to room surfaces.

It is important to note that concave reflective surfaces should be avoided for any use. If these types of surfaces must be used, it is best to provide the concave shape with acoustically transparent cloth covering absorptive or diffusive materials. If feasible, as was done in the Somerset County Courthouse, absorptive spray-on material can be used on domed ceilings. Absorptive materials should be used when a reduced reverberation time is desired in a larger room (e.g., for lectures) and diffusive materials should be used when an enhanced reverberation time is desired in a smaller room (e.g., for cathedral music).

Case Study: Flat Walls—Spivey Hall (Morrow, GA)

Spivey Hall, located on the campus of Clayton College and State University, which opened in 1991, was designed purely for music—though a wide variety of music, from organ to piano. The hall seats approximately 400 and is basically rectangular, about 50 ft (15 m) tall at its highest point, about 100 ft (30 m) long, and 56 ft (17 m) wide.

Although this may sound like a shoebox design, that was hardly the case. To maximize acoustical quality, the ceiling is pitched in a cathedral style, the rear wall is convex to diffuse sound, and the walls are stepped and sloped, as shown in Figure 3.15. The only permanent surface in the hall that is absorptive is the seating area. The seats are padded and all other surfaces are reflective to maximize the reverberation time in the room for organ recitals. The floor is wood, the ceiling is gypsum board, and the walls are part plaster on masonry and part gypsum board on randomly spaced framing. Figure 3.16 shows the interior of the hall.

One of the many unique aspects of this hall is the potential to adjust the reverberation time. This can be accomplished using the retractable curtains on the side and rear walls, which can be opened and closed at the touch of a button. Depending on the exposure of the curtains, the midfrequency reverberation times range from 1.7 to 2.4 sec.

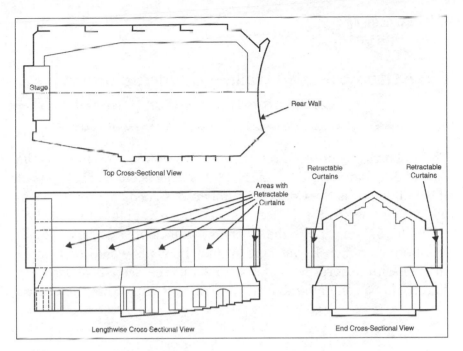

Figure 3.15 General plans for Spivey Hall.

Figure 3.16 Spivey Hall, view from the audience. *(Photograph by Rion Rizzo, Creative Sources Photography.)*

Case Study: Curved Walls—Cabarrus County Governmental Center (Concord, NC)

The Cabarrus County Governmental Center was designed in the late 1980s and built in the early 1990s. The acoustic challenge of this building was the Council Chamber, designed as a round room with a hard, reflective domed ceiling. The high point of the ceiling is 27 ft (8.2 m) above the floor and the floor area has a 42-ft (12.8-m) diameter. It was agreed that the walls would be heavily treated with absorptive material to avoid the acoustical focusing that could be generated, but the designers insisted that the domed

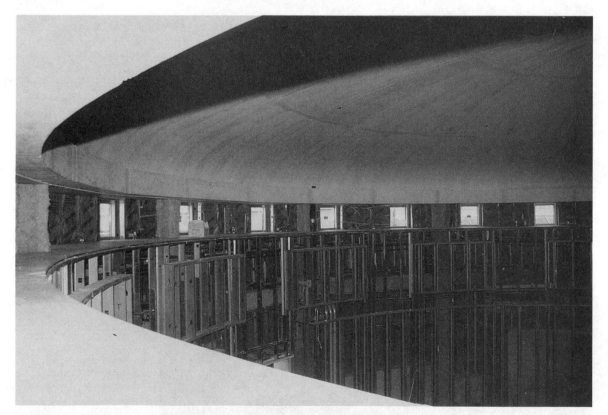

Figure 3.17   Cabarrus County Governmental Center during construction, showing the curved side walls and the dome ceiling.

ceiling remain hard and reflective. The walls were treated with
1-in (2.5-cm) fabric-wrapped glass fiber panels, backed by a large
air space containing two 10-in (25-cm) blankets of unfaced insulation. The total wall thickness is 20 in (50 cm). The floor was
covered with ½-in (1-cm) carpet on a ½-in (1-cm) pad.

With all these absorptive measures, the room (with no seats or
occupants) still had a hot spot because of the exposed reflective
domed ceiling. The conditions improved greatly with the addition
of padded pewlike benches. The absorption on the floor also minimized any flutter echo generated between the floor and the ceiling. Figures 3.17 through 3.20 show photographs of the facility
both during and after its construction.

Figure 3.18   Interior of the courtroom during construction, showing the extent of absorptive material in the side walls.

Figure 3.19 Completed courtroom, view from the side of the room. *(Photograph by Joann Sieburg-Baker.)*

Figure 3.20 Completed courtroom, view from the rear. *(Photograph by Joann Sieburg-Baker.)*

Case Study: Clouds—Kresge Auditorium (Cambridge, MA)

Many performance halls have high and curved ceilings that can cause sound to be directed away from the stage and audience or focused to create hot spots (and, consequently, zones of poor coverage). One remedy for this situation is to add a complete, properly shaped ceiling or to partially shield the curved ceiling with suspended panels. Suspended panels, often referred to as *clouds* (since they are suspended in midair above the stage and audience), are reflective to direct the unfocused sound to the audience as well as the performers on the stage. Kresge Auditorium is an example of this design, with a grapefruit-shaped roof that is also the shape of the concrete ceiling inside.

Kresge Auditorium is in the center of the Massachusetts Institute of Technology (MIT) campus in Cambridge, Massachusetts. It was meant to be a multipurpose facility, but is used mainly for lectures and drama. As described by Leo Beranek in *Music, Acoustics and Architecture* (John Wiley & Sons, 1962), this hall was "among the best in the United States" for recitals and small musical ensembles in the early 1960s. There is an organ in the hall which does not receive favorable reviews because of the low reverberation time in the hall for that type of music. The hall seats roughly 1200 and has a midfrequency reverberation time of approximately 1.5 sec, fully occupied.

As already mentioned, the original design (in the 1950s) included a concrete domed ceiling. Realizing that the dome would cause acoustic problems (especially hot and dead spots), the acoustical consultant designed suspended panels to reflect sound more evenly to the audience than would be accomplished by the domed ceiling. These plaster panels initially provided acceptable acoustics, but dissatisfaction with the acoustics was being voiced over time by many, especially the performers. The flat panels, although alleviating the effects of the dome, were not spreading sound evenly to the audience and were not offering much at all to the performers on stage (who complained of not being able to hear each other).

These problems remained uncorrected until the late 1990s, when the original flat panels were replaced with specially curved panels. The new panels were installed in the same locations as the old panels, shielding the dome and the ventilation equipment in the hall. The results have been praised by everyone from conductors to audience members. The performers say that they can now hear each other much better than before. The new panels were designed to spread the sound evenly over the audience and the stage. Their uneven shapes can be seen from the accompanying photographs of the room. Of note is that the rear wall is covered with absorptive material to eliminate echoes that can be generated from rear wall reflections. Also, the hanging panels are white and some of them have grilles through which ventilation equipment is connected. This equipment has been painted black to blend in with the dark dome ceiling and minimize its visibility to the audience. The new panels and the rest of the auditorium's features are shown in Figures 3.21 through 3.25.

Figure 3.21 Kresge Auditorium, view of new clouds from side of stage. *(Photograph by Joan McQuaid.)*

Figure 3.22 View from stage into audience, showing clouds above the stage and above the rear seating area. *(Photograph by Joan McQuaid.)*

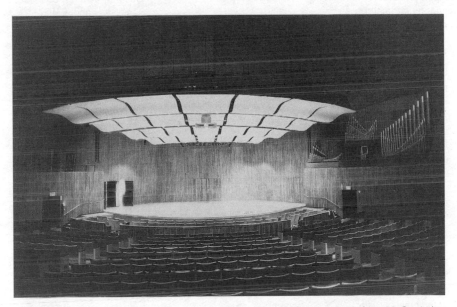

Figure 3.23 View from the audience of the stage area. *(Photograph by Joan McQuaid.)*

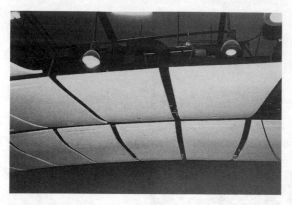

Figure 3.24   Reflecting clouds over the rear seating area, showing the integration of vents and HVAC equipment. *(Photograph by Joan McQuaid.)*

Figure 3.25   Absorptive panels on rear wall of auditorium. *(Photograph by Joan McQuaid.)*

Interior Noise Control

Design Tools

Building Component	Design Considerations	Potential Problems	Solutions
Interior walls	Appropriate TL, sealing perimeters and penetrations	Compromising privacy	Use multilayered designs; seal walls and perimeters with nonhardening materials.
Floor/ceiling assemblies	Appropriate TL and isolation, penetrations	Footfall noise, lack of privacy	Use floating floors; seal penetrations with nonhardening materials.
Wall components	Doors, windows	Compromising privacy	Use fully gasketed, multilayered doors and windows.
HVAC	Minimizing HVAC noise	Intrusive background noise	Size designs for minimum flow turbulence; insulate ductwork; isolate equipment and ductwork with resilient hangers and materials; use active noise control where feasible; locate equipment room as remote as possible from critical spaces.
Plumbing noise	Avoid rigid contact with common walls	Water rush and banging noise at remote locations	Isolate pipes within walls with resilient materials or hangers; wrap pipes with insulation.

Case Study: Noise Reduction Between Floors—
Berklee College of Music
(Boston, MA)

In 1997, the Berklee College of Music opened a new building housing classrooms, practice rooms, and a performance recital hall. This building was gutted from what had been an automobile repair facility in the 1920s and an educational building more recently. An advantage in this project was that the architect and the project team hired an acoustical consultant in the design stages. In this way, the appropriate acoustic considerations were designed into the facility, rather than patched in as an afterthought (as happens with many projects).

The key acoustical concerns with the classrooms and practice rooms were privacy and low background noise. These were achieved through designing the common wall, ceilings, windows, and doors appropriately. Common walls are of a double-wall design, with air spaces between studs and insulation in the cavity to ensure that there are no rigid connections between the sides of the walls. Windows are double-paned, having two separate layers of glass at least 3 in (7.6 cm) apart. Doors are solid-core wood and fully gasketed. These are shown in Figures 3.26 and 3.27. Sound-absorbing panels were also installed on the walls of the practice rooms. These consist of 1-in-thick (2.5-cm-thick) glass fiber covered with an acoustically transparent fabric. With a ⅛-in-thick (⅓-cm-thick) layer of a higher-density glass fiber board inserted between the fabric cover and the glass fiber core, the panels also serve as tackboards.

The other key component to achieving a low background noise level is control of the HVAC noise. This was accomplished by lining and isolating ducts, using appropriate silencers at the mechanical equipment, and minimizing airflow velocities. Ducts were lined with a 1-in (2.5-cm) thickness of absorptive material. They were isolated by using resilient hangers. One other issue with the HVAC system was the air-cooled liquid chiller (approximately 50-ton capacity) mounted on the roof of the building. This equipment was mounted on spring isolators with acoustically

Figure 3.26 Inside view of practice room in the Berklee College of Music, showing solid-core gasketed doors. *(Photograph by Joan McQuaid.)*

lined ductwork to minimize sound transmission into the building. Care needed to be taken to size the spring isolators for their proper deflections.

To minimize plumbing noise in the sensitive rooms, piping was attached to the toilet side of all double walls with resilient isolation brackets.

The recital hall involved many of the concerns that we have for performance spaces. This hall is small, seating roughly 150, with a balcony (as can be seen in Figures 3.28 and 3.29). Its intended use is for piano and vocal performances, jazz ensembles, small pop-rock bands, and cabaret. The concave shape of the rear wall necessitated absorptive treatment for these types of venues. This was accomplished by installing perforated metal sandwich panels

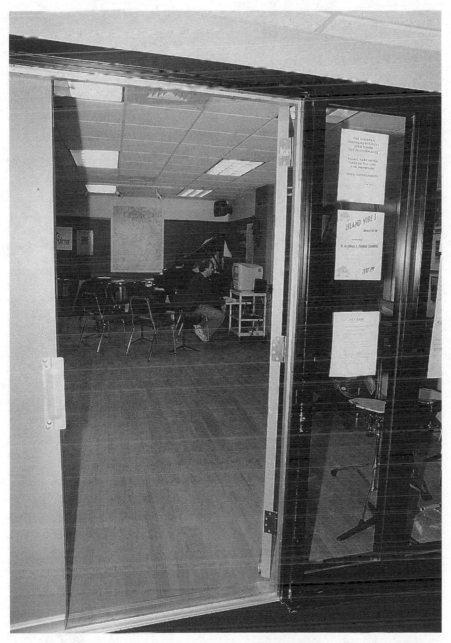

Figure 3.27 Outside view of practice room doors. *(Photograph by Joan McQuaid.)*

Figure 3.28   Recital hall at Berklee College of Music, view from the stage toward the audience. *(Photograph by Joan McQuaid.)*

housing glass fiber insulation material on the rear walls of both the lower and upper levels (see Figure 3.30). Some variation of reverberation time is possible by using retractable heavy curtains at the rear of the stage and along rear windows on the balcony (see Figure 3.31).

Figure 3.29 View of recital hall from the audience toward the stage. *(Photograph by Joan McQuaid.)*

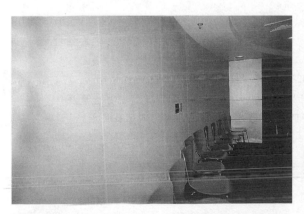

Figure 3.30 Rear walls in the recital hall, showing absorptive perforated metal panels. *(Photograph by Joan McQuaid.)*

Figure 3.31 Retractable heavy curtains used to adjust reverberation time in the recital hall. *(Photograph by Joan McQuaid.)*

Case Study: Wall Components—Airport Sound Insulation Programs

As part of its Airport Improvement Program, the Federal Aviation Administration (FAA) is funding sound insulation for residences and schools in areas around airports where 24-hour energy averaged* noise levels (called L_{dn}, or day-night equivalent noise levels) exceed 65 dBA due to aircraft operations. The goal for these programs is to provide 5 to 10 dBA of noise reduction, depending on the aircraft-generated levels, in addition to that provided by the current construction.

Standard construction materials typically provide between 20 and 25 dBA of reduction from outside sound sources. While typical wood frame exterior wall construction can provide more than 35 dBA of noise reduction from outside sources, windows and doors typically provide less than 25 dBA of reduction. We therefore look into replacing windows and doors as a first step in this insulation process. In some cases, one room of preference per dwelling is treated with additional noise reduction such as adding a layer of drywall, glass fiber insulation, and a 1-in (2.5-cm) air space to interior walls. These designs are shown schematically in Figure 3.32. This sets up a double-wall construction for the room, including adding a single layer of drywall supported on resilient channels for the ceiling (as is shown schematically in Figure 3.33).

Sometimes, central air-conditioning is installed to allow the home owners to keep their windows closed all year. Another air-conditioning option is to install an air conditioner sleeve with a wall unit and an insulated cover for winter.

From experience with sound insulation programs for more than 20 airports around the United States, amounting to specifying treatments for more than 5000 homes and testing (both before

*Note that the L_{dn} is not a straight average of noise levels over a 24-hour period; it includes a 10-dBA penalty added to all sounds occurring between 10:00 P.M. and 7:00 A.M. to account for the added sensitivity during normal sleeping hours.

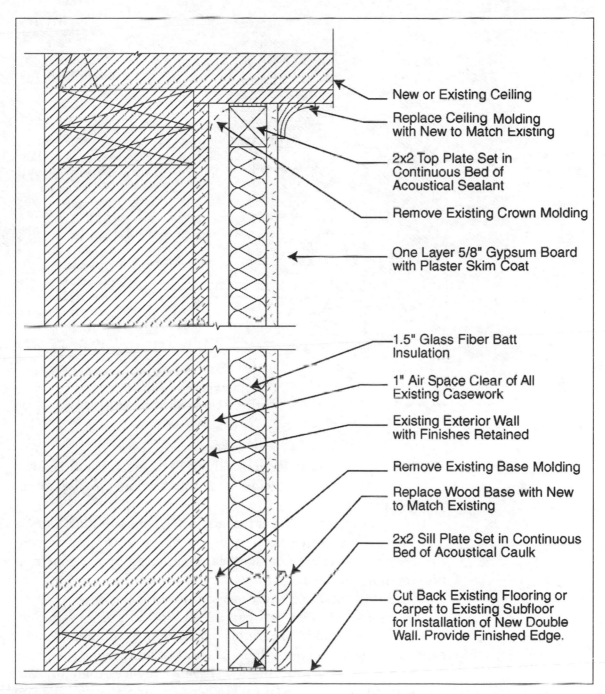

New or Existing Ceiling

Replace Ceiling Molding
with New to Match Existing

2x2 Top Plate Set in
Continuous Bed of
Acoustical Sealant

Remove Existing Crown Molding

One Layer 5/8" Gypsum Board
with Plaster Skim Coat

1.5" Glass Fiber Batt
Insulation

1" Air Space Clear of All
Existing Casework

Existing Exterior Wall
with Finishes Retained

Remove Existing Base Molding

Replace Wood Base with New
to Match Existing

2x2 Sill Plate Set in Continuous
Bed of Acoustical Caulk

Cut Back Existing Flooring or
Carpet to Existing Subfloor
for Installation of New Double
Wall. Provide Finished Edge.

Figure 3.32   Typical double-wall design for a room of preference in an airport sound insulation program.

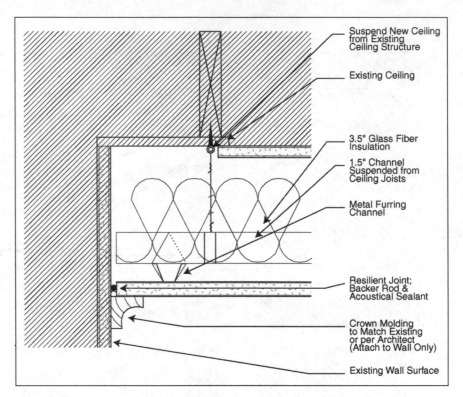

Figure 3.33 Typical double-wall design for the ceiling of a room of preference in an airport sound insulation program.

and after installations) for more than 1400 homes and schools near airports, the data included in this section provide sound reduction averages.

One point worth noting is that replacing windows and doors is typically most effective for frequencies above 250 Hz. Frequencies below 250 Hz are not significantly affected by the construction revisions in these programs.

Typical double-hung and casement exterior windows with no storm windows provide between 22 and 27 dBA of noise reduction from the outside. This noise reduction is the actual measured difference between indoor and outdoor sound pressure levels with the actual installation. It is usually less than any laboratory rating (such as STC or TL) because laboratory ratings do not consider actual installations that may not be fully sealed. These ratings also do not account for the absorption in the room in which measurements are performed. Replacement windows

Figure 3.34 Inside view of a typical double-hung replacement window used for an airport sound insulation program.

Figure 3.35 Outside view of a typical double-hung replacement window used for an airport sound insulation program.

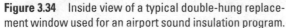

are double-paned with two windows 2 in (5 cm) apart. These are shown in Figure 3.34 from the inside and Figure 3.35 from the outside. Figure 3.36 shows a gliding replacement window. In each case, these windows offer noise reductions in the range of 33 to 38 dBA.

Typical exterior doors (shown in Figure 3.37 for a school) and sliding doors provide 26 to 29 dBA of noise reduction. Replacement doors are solid core and fully gasketed, offering noise reductions in the range of 34 to 38 dBA. Figure 3.38 shows a replacement door for a school.

On the average, the replacement of windows and doors on these houses resulted in an additional noise reduction of 9 dBA over the capabilities of standard construction materials.

Figure 3.36 Typical gliding replacement window used for an airport sound insulation program.

Figure 3.37 Typical school exterior door.

Figure 3.38 Typical replacement exterior door for a school, solid core and fully gasketed, used for airport sound insulation programs.

Case Study: Active Noise Control— Locomotive Noise

This is the only case study in this book not directly performed by Acentech personnel because they do not specify active noise control in their practice. This is because of its limited applicability and high cost, which often make it an impractical noise control option. The closest we could get was to enlist a case study from Acentech's former parent company, BBN Technologies, where many proprietary projects have been completed using active noise control technology, but few containing information available to the public.

There is little work being performed to use active noise control in buildings, with the exception of ventilating systems, industrial plants, and aircraft cabins. Active noise control headsets arc also being developed for different repeatable acoustic environments. This case study, although not directly applicable to buildings, gives a feel for the applications and effectiveness of active noise control systems.

As is mentioned in Chapter 2, active noise control is most effective in enclosed environments where dominant noise sources are low in frequency and tonal in nature. Locomotive engines provide this situation. This study was performed in 1999 for the Federal Railroad Administration on the locomotive pictured in Figure 3.39. Figure 3.40 shows the active noise control system, with microphones and loudspeakers, mounted on the roof of the locomotive. The locomotive was also fitted with a passive muffler, similar in design to those found in large HVAC systems. Since passive mufflers are effective for frequencies above 500 Hz and active mufflers are effective for frequencies below 500 Hz, this

Figure 3.39 Locomotive used in the active noise control study.

Figure 3.40   Active noise control system, with loudspeakers and microphones, mounted on top of the locomotive.

combination of methods provides an effective noise reduction design for the exhaust noise. This type of design has also been proposed and tested for automobile and bus mufflers; however, cost issues have halted their mass acceptance.

With the active system operating, tonal components below 250 Hz were reduced by up to 30 dB. These reductions provide a dramatic difference that is clearly noticeable when the system is turned off and on. Similar reductions have been seen using active systems in loud ventilation systems and industrial plants. Although there has been talk in architectural circles of the development of an active noise control system that can be part of a wall, the sources that can be controlled in this fashion will be very limited until signal processing speeds advance significantly. This may become a reality in the near future, but it is not available at the time of this writing.

CHAPTER 4

Sound Privacy/Isolation

The key acoustical issues covered in this chapter are low-frequency absorption/transmission loss, noise reduction/privacy, masking/background noise control, and multiple-unit buildings.

Low-Frequency Absorption/Transmission Loss

Design Tools

Room Purpose	Design Considerations	Potential Problems	Solutions
Indoor recreation (e.g., gymnasium or natatorium) or large atrium	Minimize reverberation.	Annoying background noise; lack of speech intelligibility	Low-frequency absorption designs
Quiet—recording studio, library, or laboratory	Allow for low background noise, significant isolation between rooms and from outdoor sound.	Background noise levels too high; inadequate privacy between rooms	Quiet HVAC designs; room isolation design

Case Study: Studio—Soundmirror Inc. (Jamaica Plain, MA)

Soundmirror is a recording studio unassumingly built in a residential community, as are many recording studios. The building is on the corner of two local streets having car and truck traffic, and a neighbor has a rather vocal dog that is in the outside yard most of each day. There are studios on two floors, one being below ground level.

There are no windows or doors to the outside in any of the studio rooms, each one of which is enclosed by heavy masonry walls that are separated from each other by air spaces and resilient materials. There is a double door to create a sound lock between studios, with each door being filled and fully gasketed. Double-paned windows are in the common wall between studio rooms to allow for communication; however, the windows are tilted to minimize the potential for standing waves. Diaphragmatic absorbers (sized to the rooms) were built into the edge of the ceilings of each

Figure 4.1 Soundmirror Inc. studio, showing fully gasketed double doors, double window with tilted panes, and diffusive panels. The diffusive panels can be replaced with absorptive panels, as needed. *(Photograph by John Urban.)*

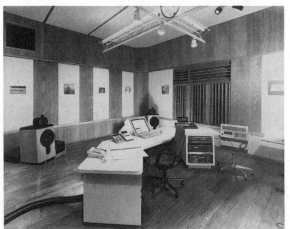

Figure 4.2 Studio at Soundmirror Inc., showing a large ventilation grille to minimize HVAC noise and removable diffusive panels. *(Photograph by John Urban.)*

studio room to minimize standing waves. HVAC background noise was minimized in the rooms by minimizing the flow velocity, lining the ducts with acoustically absorptive materials, and installing specially designed silencers tuned to the dominant frequencies of the fans.

These meticulous designs result in significant isolation between rooms, low background levels, and an ideal environment for recording. The materials on the walls are removable to set up reverberant spaces with diffusive materials or dead spaces with absorptive materials. Figures 4.1 and 4.2 show some of the interior details of the facility.

Case Study: Atrium—Worcester Art Museum (Worcester, MA)

The Worcester Art Museum was built in 1898 in downtown Worcester, Massachusetts. The lobby area, known as Stephen Salisbury Hall, was not exceptionally large but was surrounded by hard, smooth surfaces such as ceramic tile, plaster, marble, and concrete block. The ceiling was at a height of roughly 20 ft (6 m). A staircase is on one side of this lobby, where the reverberation made it difficult for people to communicate if they were not very close to each other, especially when others were conversing in the hall. As with many historic structures, and especially museums, preserving the aesthetics was a primary consideration. However, something had to be done to alleviate the acoustical situation.

The only surface that was available for treatment was the ceiling. As part of the renovation of the hall in 1997, it was decided to use spray-on cellulose material on the ceiling to provide absorption for the room. After a 1-in (2.5-cm) thickness of material was sprayed on the ceiling, there was a noticeable reduction of reverberation and improvement of the acoustics in the space. It was now possible to converse at normal levels for the first time since the hall opened. The spray-on material is also unnoticeable to visitors. Figures 4.3 through 4.6 show the room and the material on the ceiling.

Figure 4.3 Salisbury Hall in the Worcester Art Museum. *(Photograph by Joan McQuaid.)*

Figure 4.4 View toward the stairwell from Salisbury Hall. *(Photograph by Joan McQuaid.)*

Figure 4.6 Closer view of the ceiling, showing the texture of the spray-on absorptive material. *(Photograph by Joan McQuaid.)*

Figure 4.5 View of the ceiling from a distance. *(Photograph by Joan McQuaid.)*

Noise Reduction/Privacy

Design Tools

Facility	Design Considerations	Potential Problems	Solutions
A loud facility sharing a common wall with a quiet facility	Privacy, minimizing intrusive noise	Intrusive noise in quiet space	Seal all openings and perimeters; use double walls, floated floors, and resilient connections for piping and ductwork.
Restaurant	Privacy, intimacy	Harsh, loud environment; lack of privacy	Add absorption to ceilings and side walls.

NOTE: Whenever possible, plan a nonsensitive corridor between the outer wall of a noise-sensitive room and a noise-generating room.

Case Study: Yoga Facility Adjacent to Aerobics Facility

The next two categories of case studies share a common trait: their designs are so controversial that the owners typically do not wish to be identified. They involve designs of adjacent spaces that are acoustically incompatible—an inherently loud space adjacent to a space requiring quiet. Rather than avoid this type of adjacency, the owners and developers chose to design these facilities and let the acoustical consultants work out the details. Unfortunately, the constraints involved with these types of projects often set up impossible problems for the acoustical consultant to solve. It is for these reasons that these types of projects are also often involved with legal action.

The first piece of advice, then, is to avoid this type of design. However, this generic project offers the basic designs that can be used to address these problems. This project involved an unlikely adjacency—a yoga facility sharing a common wall with an aerobics facility. Aerobics facilities typically have high-powered sound systems for amplified music and voice used during aerobics classes. In the originally designed configuration, the low-frequency (bass) energy from the amplified music and midfrequency amplified voice were disruptive to yoga activities in the adjacent space. As a potential solution to this problem, it was proposed to add a second wall (extended to and sealed with the structural ceiling) with a 1-ft (30-cm) air space to the original common wall. The new second wall would consist of two layers of gypsum wallboard having staggered joints. One advantage of this project was that the space was available for this extra wall and air space. This is often not the case, but the air space is necessary to significantly contain the low-frequency sound generated in the aerobics facility. It was also recommended that a 6-in (15-cm) layer of unfaced glass fiber insulation be loosely installed in the air space between walls.

No penetrations could be introduced into the new wall, as they would compromise its noise reduction effectiveness. It was also critical that the perimeter of the new wall be sealed with a nonhardening material, such as silicone caulk or neoprene pads. As is typical with these types of projects, the client is heard from only if there is a problem with the design. Response is typical only for noncontroversial projects.

Case Study: Mechanical Room Below Quiet Offices

As for the last case, this case involves an inherently loud space adjacent to a quiet space. This is another example of the type of case study for which the participants do not wish to be identified, but the details are instructive. In this case, a mechanical room housing three compressor skids that would generate in excess of 100 dBA was planned to be on the floor below the offices of a publishing house. Measurements were taken on similar equipment at another facility, which revealed sound levels approaching 110 dBA uniformly in the room, with most of the energy around 250 Hz.

There were then two issues that had to be addressed to attempt to control the noise to the point at which the office occupants would be comfortable with their neighbors. The main issue was isolating the noise between floors, which was particularly challenging since a significant amount of low-frequency sound would be generated by the mechanical equipment. The second issue to address was the additional sound in the mechanical room caused by reverberation within the room. The latter issue was much simpler to address than the former.

The main issue of isolation between floors was addressed by first performing acoustical tests on the isolation capabilities of the existing floor. The existing floor was a concrete slab, which should provide a reasonable amount of noise reduction between floors. However, this testing revealed many holes in a concrete slab that was assumed to be solid. These penetrations were for heating, plumbing, and gas piping. Any holes in floors are clear paths through which airborne sound can be transmitted. Therefore, independent of the design of the new floor, these holes had to be sealed. With these holes properly sealed with resilient materials, the concern was structure-borne noise transmission. This is typical for low frequencies and can only be effectively addressed by isolation designs.

The only way to effectively reduce this noise transmission between floors is to install a floating floor. A floating floor is structurally isolated from a building by a resilient support system either as individual isolating pads or springs, or as a roll-out isolation grid. This system is set out, a form of plywood or steel decking is laid on the system, and concrete is poured on top to complete the flooring system. Care must also be taken to ensure that the perimeter of the floated floor is isolated from the building structure. This can be accomplished by air gaps or resilient materials. Resilient materials are recommended to seal the floor to the structure and ensure that minimal noise is transmitted. Another consideration is space. A floating floor typically adds 4 to 7 in (10 to 18 cm) to the finished elevation of the subfloor, largely dependent on the depth of the space between floor layers and the thickness of the floating concrete slab.

To lower the reverberation (and thus the sound level) in the mechanical room, it was recommended that a 2- to 3-in (5- to 8-cm) thickness of spray-on absorptive material be added to the ceiling. As with the last case study, we hear back from the client only if there is a problem with the installation. We therefore have assumed that these recommendations proved satisfactory for the situation.

Case Study: Restaurant—No. 9 Park (Boston, MA)

Restaurants often provide an environment where large groups of conversing people are compacted into small areas. Even with the best of acoustical conditions in such establishments, a localized space housing many noise sources is not an ideal recipe for acoustic privacy. There are measures that can be taken, usually limited to added absorption, to minimize acoustic problems in restaurants. This case study provides an example of the difference that can be made by minor acoustic modifications.

No. 9 Park is a small, high-end restaurant that opened in Boston in 1998. The restaurant has three separate spaces, each with its own character, and each with its own acoustic condition. Originally, each room had hard, smooth plaster ceilings at a height of approximately 9 ft (3 m). The entry room also serves as the bar and lounge. The floors are tile. The original main dining room was long and narrow, and the walls were upholstered with a green felt, backed by a ⅜-in-thick (1-cm-thick) upholstery batting. The rear dining room had all hard surfaces. None of the rooms had large tables, and the cacophony and reflected sound detracted from the elegant atmosphere of the restaurant. The environment was described by the manager as being harsh, distracting, and intrusive, and people were visibly edgy from the acoustics. Some patrons even complained that they would not return to the establishment because of the harsh environment.

Figure 4.7 Bar and seating area in the front room of the No. 9 Park restaurant, with absorptive panels on the ceiling. *(Photograph by Joan McQuaid.)*

Figure 4.8 One of the dining rooms, with absorptive panels on the ceiling. *(Photograph by Joan McQuaid.)*

As is often the case, the restaurant management wanted to solve the acoustic problem without changing the aesthetics. The tile floor was not to be touched, eliminating the possibility of carpet for absorption. They also did not want to absorb all of the sound since that would create too much of a sterile environment.

The solution used for this problem was to cover the ceilings in the three rooms with sound-absorbing panels. These panels were basically glass fiber board covered with acoustically transparent cloth to blend in with the environment. If it weren't for the seams in the panels, there would be no difference in the look of the ceiling. Figures 4.7 through 4.11 show the resulting aesthetics.

Figure 4.9 The other dining room in the restaurant, with the seams of the absorptive ceiling panels more obvious. *(Photograph by Joan McQuaid.)*

Figure 4.11 Upholstered wall in one of the dining rooms. *(Photograph by Joan McQuaid.)*

Figure 4.10 The end of a ceiling panel in a dining room. *(Photograph by Joan McQuaid.)*

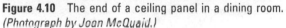

The resulting acoustics were astonishing to the management and clients. In what has been described as a "dramatic" difference, the ceiling treatment has taken the harsh edge off the acoustics of the rooms while keeping the energy desired by the restaurant management. The annoyance of the space is gone, making the front room feel more like a living room than a noisy restaurant. The other dining rooms now provide intimate environments.

Masking/Background Noise Control

Design Tools

Component	Design Considerations	Potential Problems	Solutions
Absorption	Reflective surfaces can increase annoying background noise and degrade privacy.	Excessive reverberation, hearing noises from distant areas of open office	Add absorptive materials wherever possible.
Masking	Low background sound level and intrusive noise are present.	Lack of privacy and intrusive background noise	Add electronic masking system appropriate for space.
HVAC	HVAC noise should be minimized.	Intrusive background noise	Size designs for minimum flow turbulence; insulate and isolate ductwork; use active noise control when feasible
Sound isolation	Intrusive sound from other rooms or outdoors should be minimized.	Intrusive background noise	Use vibration isolation design, resilient materials to break solid connections for structure-borne sound travel.

Case Study: Delaware County Justice Center (Muncie, IN)

This is a case in which prison cell blocks are on the second and third floors of a building that has courtrooms and offices on the first floor. Noise sources from the cell blocks, such as slamming doors, footfalls, showers running, and toilets flushing, were clearly audible in the courtrooms and offices below. Other acoustic issues in this building were excessive reverberation in the cell blocks, focusing of sound in the courtrooms because of their circular shape, and excessive mechanical noise in the circuit court. These issues were making life more difficult for prisoners and workers on the cell block levels and making courtrooms unusable on the first floor.

Standard sound isolation techniques, such as isolating pipes and ceilings with resilient hangers and wraps (as are shown in Figure 4.12), were used to reduce sound transmission between floors. Cell doors had supports installed to minimize rattling and seats and tables in cell blocks had vibration-damping materials added to their undersides to minimize their impact noise. However, these measures did not do enough to reduce the sound transmission through the building and the low background sound levels made this more obvious.

To compensate for the difference needed, electronic masking systems were installed in the courtrooms, judge's chambers, jury rooms, conference rooms, and reception and public areas. Of

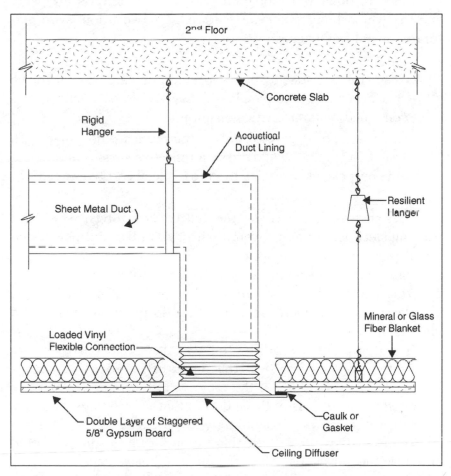

Figure 4.12 Sound isolation methods used between courtrooms and prison cells in the Delaware County Justice Center.

prime importance was raising the background levels in the frequency ranges below 500 Hz. This is because most of the remaining noise transmitted through the building was low frequency in nature, since the higher frequencies are more effectively reduced by the traditional sound isolation methods. This made the first-floor rooms usable for all occupants.

The reverberation in the cell blocks was reduced by installing perforated metal panels covering glass fiber blankets on the side walls. These metal panels expose enough of the absorptive material to effectively reduce reverberation while providing a durable covering. This type of treatment is also effective when absorption is needed in other types of rooms where durability is an issue, such as gymnasiums or subway stations. A few cautions with perforated metal sandwiches:

1. The size and spacing of the perforations are critical to the performance of the panels (so they should be specified and designed by an acoustic consultant).
2. They should be painted only if the holes are not filled with paint (since that would create a reflective surface and defeat the purpose of the holes to expose the absorptive material).

This design was also used on the cell block ceilings, where the existing steel decks were covered with glass fiber and perforated metal tack-welded to the deck.

Noise from the mechanical system was handled using standard isolation techniques, as shown in Figure 4.13. The walls of the circular courtrooms were treated with absorptive, fabric-covered glass fiber panels.

One other problem worth noting was that each courtroom has a reflective dome ceiling over the center of the room. Attorneys speaking from the center of the room would have their voices so distorted by the effects of these domes that proceedings were disturbed. A circular wooden railing was proposed to be placed in the center of the room. With this railing in place, no one could stand in the center of the room directly below each dome. At worst, attorneys would lean on the railing, but they would be kept away from the area in which their voices would become distorted.

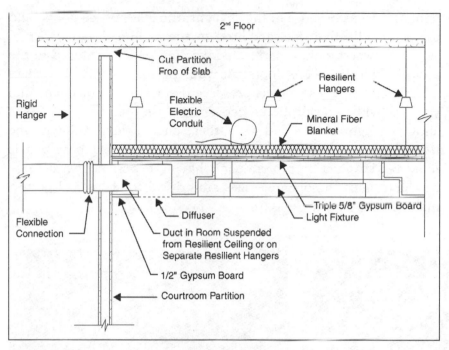

Figure 4.13 Mechanical system noise control and isolation designs between floors and between courtrooms.

Case Study: Hasbro Executive Offices (Pawtucket, RI)

In 1997, Hasbro transformed a former supermarket adjacent to its corporate headquarters into an elegant and contemporary executive office facility. The project included an atrium with a 13-sided skylight dome surrounded by private executive offices. One purpose of the atrium is to provide an area for relaxation and contemplation. Figure 4.14 shows this area after its completion.

To minimize any distractions from noise in this atrium, HVAC and mechanical system noise was minimized. Annoyance from reverberation in the space was also considered; however, the surface area available for absorptive treatment was limited to the wooden band between the tops of office doors and the domed skylight. This was because of the skylight, the glass office doors, and the hard reflective floor required in the atrium design. To preserve the wood look, a perforated wooden panel was used in the areas above the office doors. As with perforated metal used in other facilities, these panels preserve the aesthetics of the atrium while providing absorption by covering glass fiber blankets. As is shown in Figures 4.14 and 4.15, these panels blend well with the rest of the wall surfaces. As with perforated metal, these panels must be designed by professionals having experience with these products since the opening sizes are critical to acoustical performance.

Figure 4.14 Central atrium surrounded by executive offices at Hasbro's headquarters.

Figure 4.15 Closer view of side walls with absorptive perforated wood panels and open grilles for the masking and ventilation systems.

The other aspect of this project that placed it in this section of the book is the need for a sound-masking system to provide the most appropriate background noise level for the intended use of the space. The ventilation grilles in the side walls above the office doors (shown in Figure 4.15) serve a dual purpose in that respect. Part of the grilles is actually used for the ventilation system and part is covering some of the sound-masking loudspeakers. Additional sound-masking loudspeakers are hanging from the top of the skylight dome facing upward to provide diffusion of the masking sound off the dome and into the atrium.

This design package has provided the acoustical environment envisioned by Hasbro's management in its conceptual process—a space for meeting and contemplation—while not compromising the privacy of the executive offices.

Case Study: Harley-Davidson Motor Company (Milwaukee, WI)

In 1997, Harley-Davidson undertook an unconventional renovation of the first floor of an early 1900s factory building. In space that had once been used to manufacture motorcycles, Harley-Davidson and its architects devised an open-plan office environment that left most of the building structure and mechanical systems exposed. The renovated space is shown in Figure 4.16. Sound-absorbing materials are in traditional suspended ceilings (as shown in Figure 4.17), surface-mounted on the ceilings below the factory deck (as shown in Figure 4.18), and in suspended cloud sections (as shown in Figure 4.19). As is most often the case in open-plan offices, this absorption was not sufficient to provide for an adequate degree of speech privacy between cubicles since background noise levels were low. A sound-masking system was installed in the space, incorporating surface-mounted loudspeakers in perforated metal enclosures to blend in with the aesthetics. These loudspeakers can be seen below the surface-mounted ceiling panels on the structural beams shown in Figure 4.20, but are not visible behind the suspended ceilings.

Figure 4.16   Harley-Davidson offices' main lobby. *(Photograph by Ryerson Aircraft, Inc.)*

Figure 4.17   General open-office layout with traditional suspended absorptive ceiling. *(Photograph by Ryerson Aircraft, Inc.)*

Figure 4.18 Office area with ceiling-mounted absorptive panels. *(Photograph by Ryerson Aircraft, Inc.)*

Figure 4.19 Office area with suspended cloud absorptive panels. *(Photograph by Ryerson Aircraft, Inc.)*

Figure 4.20 Area with ceiling-mounted panels and vertical cylindrical masking system loudspeakers encased in painted perforated metal. *(Photograph by Ryerson Aircraft, Inc.)*

Multiple-Unit Buildings

Design Tools

Issue	Design Considerations	Potential Problems	Solutions
Privacy	Allow for acoustic separation.	Lack of privacy; annoying environment	Use isolation design of party walls including staggered studs, air spaces, and resilient materials to prevent vibration channels between units; avoid adjacent uses having conflicting loud and quiet requirements.
Background sound	Set for maximum allowable background sound to avoid privacy issues.	Lack of privacy; annoying environment	Use isolation design of party walls when low background sound levels are required (as in theaters, residences, or studios) and add masking when higher background levels are acceptable (e.g., in offices).

Case Study: Privacy Between Dwelling Units— 56 Clarendon Street (Boston, MA)

Figure 4.21 Entrance of 56 Clarendon multifamily units. *(Photograph by Joan McQuaid.)*

This is an example of what *not* to do when designing and constructing multifamily dwelling units. A three-story brick commercial structure was renovated into four condominiums, dividing the building in half with a common wall. The two upper-level floors on each side of the building became two units and the lower-level floor became two units. The exterior of the renovated building is shown in Figures 4.21 and 4.22.

Everything was fine until the owners moved in. There were two significant deficiencies with the design of these units that caused a severe lack of acoustic privacy. One was that the framing for the floor of the upper-level units crossed through the party wall. This caused any impact or footfall noise on the floor on one side of the wall to be transferred efficiently to the other side. The other design deficiency was associated with the stairs between the second and third floors within the upper level units. The stairs in each unit were designed to be next to the party wall, with the stringer boards rigidly attached to the single-stud construction of the party wall. This ensured the efficient transmission of all footfall noise on the stairs from one unit to the neighboring unit. A logical solution to this problem would have been to add a separate staggered stud in the party wall to create a double-wall construction; however, because of the locations of the stairs against this common wall, there is no space to do this without removing the stairs.

The result of these designs is that the residents can hear conversations, television, music, and footsteps from their neighbors on the other side of the party wall. If one neighbor's television is playing, the family next door can mask the sound by either playing their own television or opening their windows to introduce urban noise. However, if one neighbor's sound system is playing (with a powerful subwoofer to amplify low frequencies), there is no relief for the other neighbor.

This should be a lesson to those involved in renovating spaces into multiple residences or areas where privacy is required. Com-

Figure 4.22 Upper floors of the residential building. *(Photograph by Joan McQuaid.)*

mon walls must provide effective isolation from neighboring units, and these principles must be incorporated into designs before any construction takes place.

Case Study: Cinemaplex—Hoyts Cinema (Schenectady, NY)

The pattern of emerging cinemaplexes (housing many theaters in one building) in the United States and abroad is causing a growing construction issue of acoustic separation between movie theaters. This becomes especially challenging with the powerful sound systems being installed in these complexes. It often occurs that a movie having loud explosions is on the other side of a common wall from a theater featuring a quiet love story. Low-frequency explosions are particularly difficult to isolate using a single wood-frame construction wall. Ample air spaces between wall sections, along with isolation design, are required to provide adequate acoustic separation.

In terms of effective designs, Lucasfilm's THX® criteria offer good guidelines for acoustical separation; however, these guidelines include concrete masonry that may be impractical for some theater constructions. Based on experience designing Hoyts Cinema facilities throughout the United States and South America, Figure 4.23 offers general design guidelines for common walls between theaters when concrete is not an option (although concrete is preferable for low-frequency isolation). Even with concrete, however, air spaces are critical in the design. The total thickness of these common walls should be at least 22 in (56 cm) to effectively reduce low-frequency transmission between the-

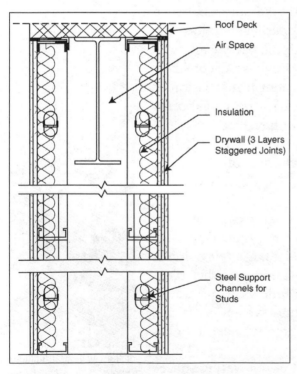

Roof Deck

Air Space

Insulation

Drywall (3 Layers Staggered Joints)

Steel Support Channels for Studs

Figure 4.23 Typical cross section of wall between Hoyts theaters.

aters. Note on Figure 4.23 that the layers of drywall should have staggered joints and all connections between the common wall and the floor or ceiling should be sealed and isolated using resilient pads or nonhardening caulking.

One design involved theaters on the floor below a shopping mall. Because of the high background noise levels in the mall and the low background noise levels required for the theaters, the noise issues concerned only sound penetrating from the mall into the theaters and not the other way around. Also at issue was foot-fall impact noise, so isolation involving resilient hangers supporting the theaters' ceilings was required (as shown in Figure 4.24).

Stadium seating is becoming popular in theater design. In one case, rest rooms were designed below the stadium seating area of some theaters. The main concern in this case was the plumbing noise from flushing toilets, which was isolated using the design shown in Figure 4.25, incorporating resilient hangers.

Important acoustic issues in cinema design other than separation are background sound level and reverberation. It is generally agreed that background sound levels should be in the NC-30 range. NC values are still used by many consultants and references, even though these criteria have been updated by NCB and RC ratings, as is noted in Chapter 2. Recommended reverberation times for cinemas vary with the room size, but they range from 0.2 sec for 1000 ft^3 (28 m^3) to 1 sec for 400,000 ft^3 (11,325 m^3).

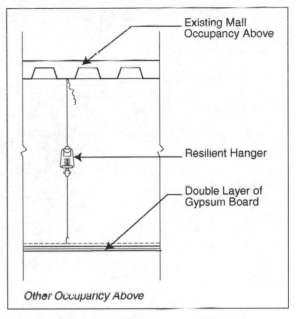

Figure 4.24 Isolation design for theater below shopping mall

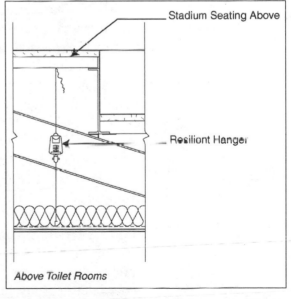

Figure 4.25 Isolation design for stadium-seating theater above rest rooms.

Case Study: Condominiums Above
Performance Space

This is another in the category of projects in which the owners did not wish to be identified; however, this general class of designs is surprisingly common in large cities. There are many buildings that house nightclubs on the bottom floors and residential units above. For one such building in New York City, the nightclub opened at midnight and generated noise and vibrations that shook the beds of residents up to five floors above the facility until 5 A.M. each night of the week. These types of adjacencies are the result of poor planning and can only be resolved by designs similar to those incorporated in the cinema houses referenced previously.

The two most significant restrictions to acoustic separation between these uses are space and common structural members. There is often insufficient space available to provide the air spaces necessary for the required isolation. Also, even with appropriate designs to isolate the ceilings of one floor from the floors below, sound (and especially low-frequency sound) can travel through connected side walls and structural members to remote building locations. This sound travel can be eliminated only by breaking the connection between structural members, which is often a prohibitively expensive task after a building has been constructed.

So the first rule in this situation is to avoid constructing facilities adjacent to each other when one can generate high sound levels and the other requires low sound levels. At a minimum, a buffer corridor between the two uses is necessary for complete isolation—that is, assuming that structural members for each use are isolated from each other by air spaces or resilient materials.

In general, it is much more difficult to isolate units in a wood-frame building than in one designed with concrete slabs and block. Even concrete slabs, however, will transmit low-frequency noise. Therefore, concrete slab floors can be floated on springs and resilient pads to provide significant low-frequency sound isolation. Floated floor construction is described in the previous generic case study for a mechanical room below an office space. Most of the time, floated floors are systems that need to be designed before a building is constructed to be practical.

CHAPTER 5

Open-Plan Offices

The open-plan office design has become very popular through the latter half of the twentieth century and into the twenty-first. This design scheme saves money, promotes teamwork, and improves flexibility for future renovations. Many employees, however, view this design as a series of compromises in terms of space, prestige, and (most of all) privacy. As employees consider changing from the traditional closed office to open-plan cubicles, they often have concerns about their abilities to work productively in what they anticipate to be a noisier, more distracting workplace. The overwhelmingly largest complaint about the open-plan office design is the lack of acoustic privacy.

This chapter has been devoted to open-plan office design to concentrate on key factors to consider for maximizing acoustic privacy in these environments.

Acoustic Considerations in Open-Plan Office Design

The first step in the acoustic design of an office is to determine the needs of the employees in each area. Employees in some companies may need to communicate freely as part of their jobs. These workers would not need any acoustic privacy. Some may need visual privacy (using barriers that block their line of sight to others) but need only a minimal amount of speech privacy. Some others may require an environment free from distractions for the performance of detailed work. It is often this last category of employees that feels most compromised by being moved into cubicles. This category of employees needs what is typically referred to as normal speech privacy, in which conversations in adjacent areas can be understood but are not distracting to concentration on tasks. Normal speech privacy is attainable for open-plan offices with the proper acoustic design.

Confidential speech privacy, in which no part of a conversation can be understood from an adjacent space, cannot be expected from open-plan office designs. If this environment is required, employees must be located in closed offices with doors and walls that extend to the *structural* ceiling. This point needs to be stressed because many closed offices are designed with doors and walls extending to suspended, nonstructural ceilings, allowing space for HVAC and electrical equipment. Typical suspended ceilings provide acoustic absorption for reflected sound but provide little in the way of transmission loss. This means that sound will travel through suspended ceiling panels and over walls as if the walls were barriers, thus limiting the acoustic privacy of these offices almost to the open-plan condition.

Speech privacy between two rooms, in general, is a function of two key parameters: noise reduction and background noise level. One must not only reduce the sound level of the unwanted source, but also generate sufficient background noise to mask the unwanted source, making it less intelligible and distracting. Confidential speech privacy can be achieved in closed offices by using walls having a high noise reduction capability.* Using these walls, the background sound levels can remain low. However, for normal speech privacy in an open-plan office, one must strike a balance between a lower noise reduction capability (because of the limited effectiveness of barriers, as mentioned in Chapter 2) and a higher background noise level (that can often be set using electronic masking systems).

Assuming a goal of normal speech privacy, there are three acoustic design principles that must be addressed to achieve that goal in an open-plan office. These three principles have been coined over the years as the ABCs (standing for *absorb, block,* and *cover*) of good office acoustics as a mnemonic device. These three principles are briefly discussed as follows.

Absorption

It is critical to have as much absorption as possible in rooms designed for open-plan offices. This will stop sound from traveling appreciable distances within rooms and from causing remote distractions. If one has to choose surfaces for absorptive treatment, however, the ceiling is the most critical surface for that treatment. This is because reflections off ceilings significantly compromise the already limited noise reduction capabilities of barriers. Ceiling finishes should have NRC ratings of at least 0.85. Ceiling treatments can be part of suspended ceilings (as shown in Figures 5.1 and 5.2) or surface-applied (as shown in Figure 5.3). Hard, sound-

---

*Note that the noise reduction effectiveness of partitions in offices is often described in terms of a parameter known as *noise reduction* (NR). Although NR is related to TL, it is not always the same as TL because NR takes the absorption of room surfaces into account, while TL does not.

Figure 5.1   Typical suspended ceiling with absorptive tiles.

reflecting lighting fixtures can degrade the performance of the ceiling finish, so it is ideal to use indirect floor lighting to keep the ceiling as absorbing as possible. If ceiling fixtures must be used, parabolic fixtures are more appropriate than others because they diffuse the sound more evenly. This and other lighting alternatives are shown in Figure 5.4*a* through *e*. Absorptive finishes are also important on

Figure 5.2   Suspended ceiling clouds.

Figure 5.3   Surface-applied absorptive ceiling treatment.

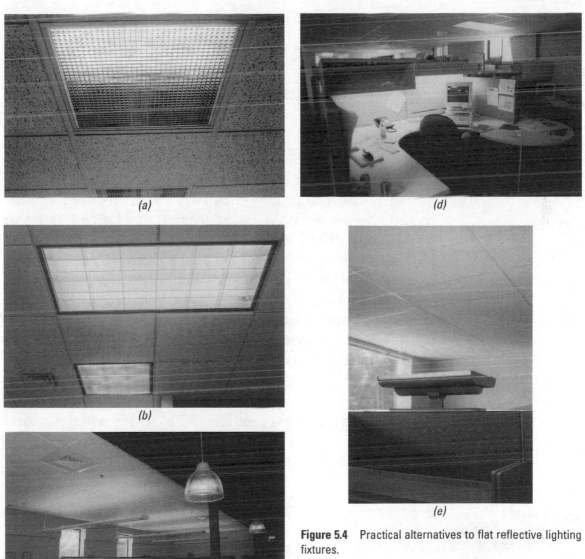

Figure 5.4 Practical alternatives to flat reflective lighting fixtures.

the interior of workstation panels and on any walls that may reflect sound from one cubicle to another. Carpeting is important for floors, more to control impact noise (such as footfalls, heel clicks, dropped objects, and furniture movement) than to absorb reflected sound energy.

Blocking

Sound is blocked in open-plan offices by cubicle barriers. To be most effective, these barriers must be designed to minimize the compromises caused by diffraction. In this regard, barriers should be at least 5 ft (1.5 m) tall and should have STC ratings of at least 25. The amount that STC ratings exceed 25 will not make a difference in barrier performance (because of diffraction). They should be arranged to block the line of sight between workers and should be flush with floors, windowsills, and side walls. Figures 5.5 through 5.7 show examples of inappropriate barrier arrangements.

Cover

A typical contemporary office, with sealed windows and properly maintained HVAC systems, often has too low of a background noise level to allow for normal speech privacy.

Figure 5.5 Inappropriate design—sound-reflective wall opposite cubicle openings.

Figure 5.6 Inappropriate design—partitions too low and narrow with gaps.

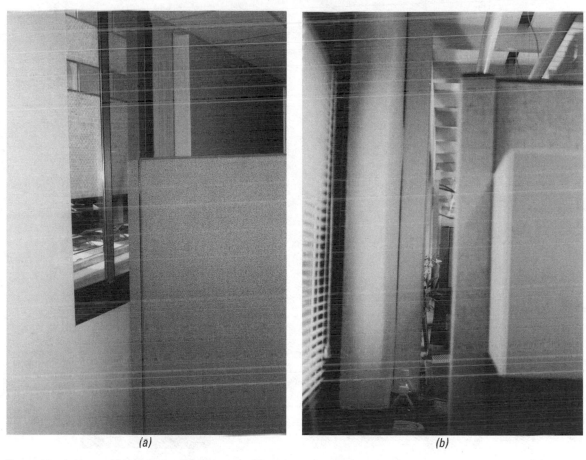

(a)   *(b)*

Figure 5.7 Inappropriate design—gaps along partition edges.

As illogical as it may sound to some, the most practical way to improve privacy under these conditions is to increase the background noise level by adding an electronic sound-masking system. These systems can include loudspeakers distributed throughout ceiling plenums, hanging unobtrusively from ceilings, or in each cubicle (as shown in Figures 5.8 through 5.11). As mentioned in Chapter 2, these systems, when set up properly, sound like typical HVAC noise to employees.

Case studies of specific designs are included following the "Design Tools" section.

Figure 5.8 Sound-masking loudspeaker above suspended ceiling.

Figure 5.9 Unducted return air grille for placement of sound-masking loudspeaker.

Figure 5.10 Surface-mounted sound-masking loudspeaker.

(a)   (b)

Figure 5.11 In-cubicle sound-masking system.

Design Tools for Normal Speech Privacy*

Design Component	Design Considerations	Potential Problems	Solutions
Absorption	Minimize reflective surfaces in open offices.	Excessive reverberation, hearing noises from distant and adjacent areas of open office	Add absorptive materials wherever possible, especially on ceiling.
Barriers	Block sound transmission for acoustic and visual privacy.	Compromise in already limited noise reduction effectiveness	Allow no air gaps; use minimum STC 25 materials; make barriers at least 5 ft (1.5 m) tall.
Masking	Account for low background sound level and intrusive noise.	Lack of privacy and intrusive background noise	Add electronic masking system appropriate for space.
HVAC	Minimize HVAC noise.	Intrusive background noise	Size designs for minimum flow turbulence; insulate ductwork; use silencers and active noise control when feasible.

Case Study: Aetna Life and Casualty Company Home Office (Hartford, CT)

In the mid-1980s, Aetna undertook a major renovation project for over 1 million ft² (93,000 m²) of office space in its headquarters. Large areas of open-plan offices were developed, allowing the possibility of testing different configurations and materials to offer the best design in terms of speech privacy. In this direction, a mock-up area was built, in which these different configurations could be evaluated. A summary of the results of this testing provides valuable guidelines for open-plan office design.

To evaluate the different designs, the *articulation index* (AI) was used. AI is a measure of speech intelligibility which ranges from 0 to 1, where 0 corresponds to no speech intelligibility and 1 corresponds to full speech intelligibility. When used to rate speech privacy between open–office plan cubicles, AI is generally a function of noise reduction and background sound pressure level.

*Note that confidential speech privacy requires closed office design with walls extending to structural ceilings.

Figure 5.12 shows how AI relates to speech privacy. Although more complicated methods exist for analyzing AI, this graph provides general guidelines from experience. Speech privacy goals were divided into three categories: minimal distraction, normal speech privacy, and confidential speech privacy. Minimal distraction corresponds to an AI of 0.35 or less. Normal speech privacy, in which the average person can work without distraction although occasional parts of outside conversations can be heard, corresponds to an AI of 0.20 or less. Confidential privacy, where the average person can carry on discussions with assurance that he or she will not be understood by neighbors, corresponds to an AI of 0.05 or less. As is mentioned in the previous section, confidential privacy is not possible in an open-plan design.

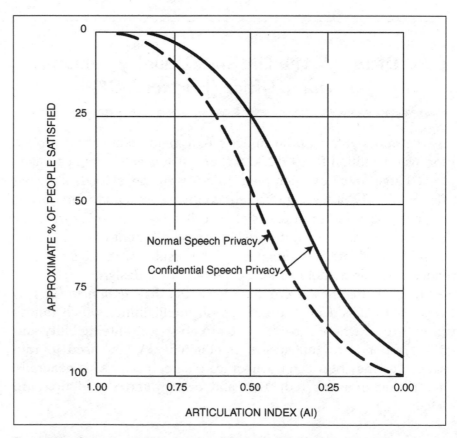

Figure 5.12 General correlation between articulation index and speech privacy.

The general conclusions from this study were as follows:

1. Barriers that are 53 in (1.35 m) tall, combined with an electronic sound-masking system, are adequate for minimal distraction.

2. Barriers that are 65 in (1.65 m) tall, combined with an electronic sound-masking system, are adequate for normal speech privacy.

3. Cubicle layouts that allow a direct line of sight between occupants will not result in normal speech privacy.

4. Side wall and ceiling reflections can degrade speech privacy and must be treated with sound-absorbing wall panels wherever normal speech privacy is required.

5. Normal speech privacy is not possible in an open-plan office arrangement without a sound-masking system, unless background levels from acceptable sound sources are already high.

Figures 5.13 through 5.16 show some of the open-plan configurations tested, along with their results in terms of AI both with and without sound masking. All cases show that sound masking is required for any type of speech privacy. Figure 5.13a and b shows that an absorptive ceiling is more effective for speech privacy than a drywall ceiling. It also shows that a 53-in-tall (1.35-m-tall) barrier is not conducive to normal speech privacy for adjacent cubicles. Figure 5.14a shows the lack of privacy with a clear line of sight between cubicles, while Figure 5.14b shows the effectiveness of a 65-in-tall (1.65-m-tall) barrier. Figure 5.14b and Fig-

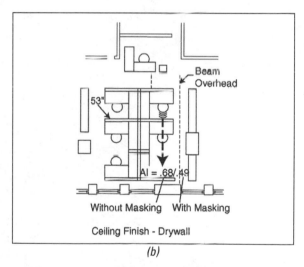

Figure 5.13 Open-office sound privacy tests with 53-in-tall (1.35-m-tall) partition and varying ceiling conditions: (a) absorptive tile ceiling, and (b) gypsum board ceiling.

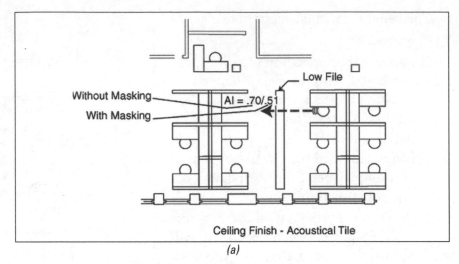

(a)

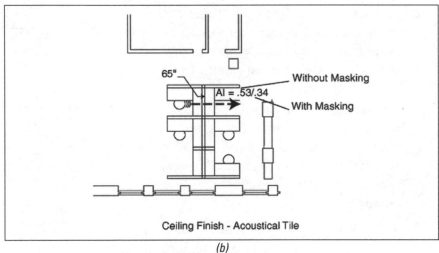

(b)

Figure 5.14   Open-office sound privacy tests with an absorptive tile ceiling showing the effect of (*a*) a filing cabinet versus (*b*) a 65-in (1.65-m) partition.

ure 5.13*a* offer a comparison with the lower barrier. Although Figure 5.15*a* and *b* adds reflective side walls, the walls introduce small AI changes from the conditions of Figure 5.14*b,* in which there are no nearby reflective side walls. In each case, however, there is an absorptive barrier between the sound source and the reflective side wall. Figure 5.16*a* and *b* shows

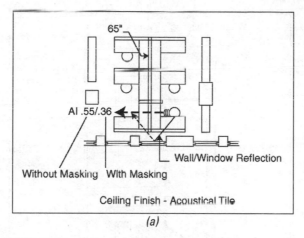

(a)

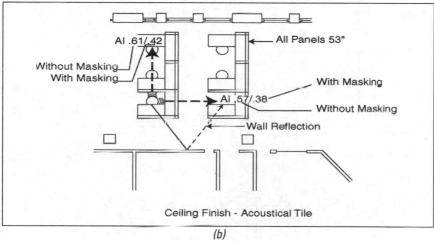

(b)

Figure 5.15 Open-office sound privacy tests with reflective side walls.

the effects of distance and multiple barriers between a sound source and listener.

Figure 5.17 shows the results of AI tests in some closed offices. As is shown in the figure, a sound-masking system is still required for confidential privacy in this study. This is especially the case when party walls terminate at suspended, rather than structural, ceilings. The effect of the unsealed glass wall can also be seen to significantly degrade the performance of the one common wall.

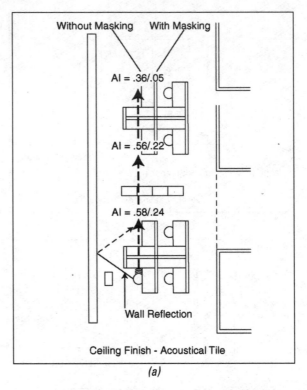

Without Masking   With Masking

AI = .36/.05

AI = .56/.22

AI = .58/.24

Wall Reflection

Ceiling Finish - Acoustical Tile

(a)

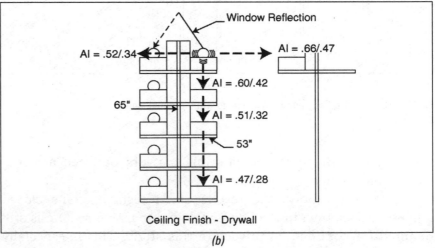

Window Reflection

AI = .52/.34   AI = .66/.47

AI = .60/.42

65"   AI = .51/.32

53"

AI = .47/.28

Ceiling Finish - Drywall

(b)

Figure 5.16 Open-office sound privacy tests with distance, with an absorptive ceiling and different configurations.

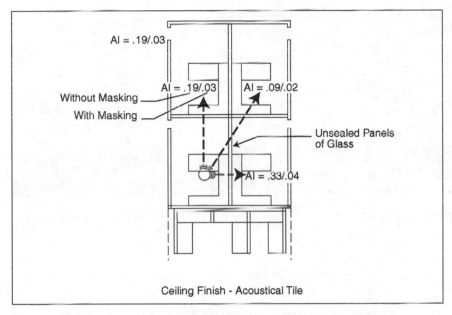

Ceiling Finish - Acoustical Tile

Figure 5.17 Closed office speech privacy assessment in Aetna's Home Office.

Case Study: MIT Career Services Center (Cambridge, MA)

The Massachusetts Institute of Technology (MIT) renovated its Career Services Center in 1995. One of the most significant concerns was acoustical privacy between stations in the interview area. The configuration was a series of closed offices, each containing two to three stations. The stations were separated by thin partitions and the side walls were painted drywall. There was little if any privacy between these stations.

The new design was a completely open plan. To provide some privacy between workstations in this design, it was necessary to incorporate generous use of absorptive treatments. Figure 5.18 shows an overall view of the renovated interviewing area. The floors were carpeted to minimize impact and footfall noise, the ceiling was covered with a suspended tile having a minimum NRC of 0.90, and the walls in each cubicle were treated with absorp-

Figure 5.18 MIT Career Services Center, view of overall facility. *(Photograph by Joan McQuaid.)*

Figure 5.19 One cubicle with absorptive panels on walls. *(Photograph by Joan McQuaid.)*

tive panels (see Figure 5.19). The partitions between workstations are absorptive themselves, with NRC values of more than 0.75. In order to maximize the absorption of the ceiling, no lighting fixtures were built into the suspended panels. Instead, lighting was provided by hanging fixtures and fixtures built into soffits, as shown in Figure 5.20.

The overall layout of the area considered blocking any line of sight between occupants of cubicles to maximize the noise reduction of the partitions. The high amount of absorption and blockage of line of sight provided the best acoustical conditions possible

Figure 5.20 Indirect lighting from hanging fixtures and fixtures in the ledge of the suspended ceiling. *(Photograph by Joan McQuaid.)*

Figure 5.21 Masking system loudspeakers behind unducted ventilation grille. *(Photograph by Joan McQuaid.)*

with this arrangement; however, the background sound level was too low to ensure adequate privacy. The final step in the design then was to raise the background level using a sound-masking system having loudspeakers hidden above the suspended ceiling. In this case, loudspeakers for the masking system were placed behind unducted ventilation grilles, as shown in Figure 5.21.

The result of this process is adequate normal speech privacy in these interviewing stations, a condition that did not exist in the former closed-office setup. When one walks through this room, conversations can be heard only when passing open sections of cubicles.

Multipurpose Facilities

Multipurpose facilities, by definition, are used for a wide variety of events (such as classical and rock concerts and drama and sporting events), all in the same room. As we have discussed, each of these types of events requires a different set of acoustic criteria. Other issues that must be considered in multipurpose facilities are background noise levels, room treatments to eliminate echoes, and electronic enhancement of sound. These issues are exemplified in the case studies of this chapter.

Design Tools

Acoustic Issue	Design Considerations	Potential Problems	Solutions
Appropriate reverberation time for the intended use	Use low RT_{60} for drama, lectures, and sporting events; higher RT_{60} for classical concerts. (See Chapter 3 for more details.)	Speech intelligibility problems; room not live enough for music	Use adjustable absorption in room and/or electronic enhancement with sound systems.

Acoustic Issue	Design Considerations	Potential Problems	Solutions
Room treatments	Avoid room designs that can cause echoes or other anomalies (e.g., reflective surfaces that are concave or flat and parallel to each other).	Echoes that can negatively affect sound quality in a room	Either eliminate counterproductive wall shapes or treat these surfaces with absorption or diffusion (depending on the need for high or low RT_{60}).
Background noise	Minimize background noise.	Excessive noise from HVAC or exterior sources that make background noise levels unacceptable	Use effective noise control measures at the source and/or in the building design.

Figure 6.1 Lowell Memorial Auditorium, view from the audience toward the stage. *(Photograph by Joan McQuaid.)*

Case Study: A Hall for All Uses—Lowell Memorial Auditorium (Lowell, MA)

The Lowell Memorial Auditorium, shown in Figures 6.1 and 6.2, was built in the 1920s and, since that time, has hosted events from the New England Golden Gloves Boxing Championships to concerts by the Boston Pops. Within that range were performances from the Three Stooges to Bruce Springsteen, from Broadway shows to trade shows. This 3000-seat auditorium is a true multipurpose facility. The floor area is under constant revision to accommodate each visiting act. It sees more diversity in performances than most arenas.

Despite initial claims to the contrary, the shape of the facility does not lend itself to the best acoustics. It has an egg-shaped interior and a domed ceiling (see Figures 6.3 and 6.4). As has been shown in other case studies in this book, these concave shapes need to be treated to avoid echoes and sound focusing. There is also a rather deep seating section under the balcony that loses much of the ambiance typical of the rest of the space.

The facility fell into disrepair and was renovated in the mid-1980s to prepare it for a significant expansion to its schedule of events. As part of this renovation, consultants were directed to address the acoustics that, up to that time, had been too live for most uses. Although many recommendations were made, the ones that

Figure 6.2 View from the stage toward the audience. *(Photograph by Joan McQuaid.)*

were adopted involved adding absorptive materials to the curved walls on both the lower and balcony levels. One key issue with the management of the facility was that the aesthetics not be changed. On the lower level, fabric-covered glass fiber panels line the curved walls. On the balcony level, perforated metal panels cover glass fiber blankets to produce the desired absorption (see

Figure 6.3 View from balcony, showing the egg-shaped nature of the facility. *(Photograph by Joan McQuaid.)*

Figure 6.4 Domed ceiling of the facility. *(Photograph by Joan McQuaid.)*

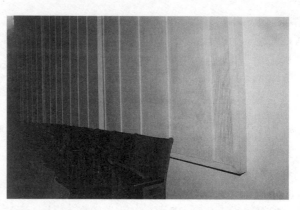

Figure 6.5 Absorptive perforated metal panels along the back wall of the balcony level. *(Photograph by Joan McQuaid.)*

Figure 6.5). In each case, the absorptive panels blend in with the room. The lower-level panels are covered with fabric that is the same color as the walls and the balcony-level panels are painted the same color as the walls. Although the room is still quite live because of all the exposed reflective surfaces and the reflective dome, echoes and focusing have been significantly reduced with the absorptive treatments. The upholstery on the main-floor seats helps to deaden reflections off the domed ceiling, though it does not entirely eliminate the dome's effects.

The underside of the balcony is reinforced by two sound systems, one distributed system with loudspeakers hanging from under the outer ledge of the balcony (as is shown in Figure 6.6) and another distributed system consisting of ceiling-mounted loudspeakers under the balcony.

Figure 6.6   Underside of the balcony, showing one of the distributed loudspeaker systems. *(Photograph by Joan McQuaid.)*

Case Study: Electronic Enhancement—
51 Walden Street (Concord, MA)

51 Walden Street is a theater used for drama, the local concert band, and the local symphony orchestra. It is housed in a building that was a cavalry practice building during the American Revolutionary War. The seating for this theater (roughly 400) faces toward the large stage for drama and toward the shell on the opposite side for symphony concerts (see Figures 6.7 through 6.9). When the facility was originally renovated to house performing arts, the reverberation time was too long for drama and too short for a symphony orchestra. Satisfying the requirements of both of these venues was accomplished by adding enough absorption to the walls to lower the reverberation time to one appropriate for drama and by adding an artificial reverberation sound system to be used for orchestral and band concerts.

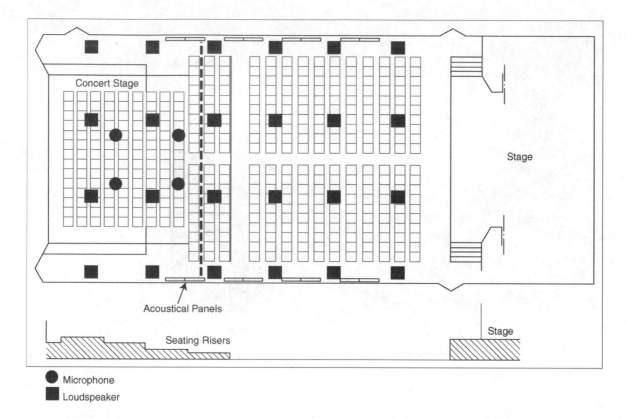

Microphone
Loudspeaker

Figure 6.7 General layout of the 51 Walden Street theater, in plan view and cross section.

Figure 6.8 View of the drama stage from the concert stage area, with seating set up for drama. *(Photograph by Joan McQuaid.)*

Figure 6.9 View of the concert stage from the drama stage, with seating set up for drama. *(Photograph by Joan McQuaid.)*

The side walls have glass fiber covered with acoustically transparent fabric in rotatable panels that can expose either the absorptive side or a reflective side. Below these panels are diaphragmatic absorbers, in the form of light wood covering an air space. To the eye, they look like cabinets built into the side walls (see Figure 6.10) and were originally built to conceal heating equipment. It was realized that these sections were functioning as diaphragmatic absorbers only when reverberation time measurements revealed a decreased RT_{60} at low frequencies, a situation that usually does not occur in rooms this size. The rear wall has wood fiber absorptive panels and a row of wooden panels tilted at different angles to provide diffusion (shown in Figure 6.11). Above the rear section are hanging panels that can be rotated to expose either absorptive or reflective surfaces. It was found that the performers could communicate much more effectively if the absorptive panels were facing toward them, so this is the preferred configuration.

Figure 6.10 Side walls with panel absorbers mounted above diaphragmatic absorbers. *(Photograph by Joan McQuaid.)*

The artificial reverberation system consists of microphones to sense the sound on the stage, signal processors that add reverberation to those sounds, and loudspeakers that reproduce this sound to the audience. Figures 6.7 and 6.12 show how these loudspeakers are spread throughout the audience area and hang from support beams. To avoid any distortion caused by sound arrival from different loudspeakers, time delays were programmed into the system.

Figure 6.11 Rear wall of the concert stage with absorptive panels above and diffusive panels below. *(Photograph by Joan McQuaid.)*

Figure 6.12 Artificial reverberation loudspeakers hung from support beams in the theater. *(Photograph by Joan McQuaid.)*

Case Study: Adjustable Absorption— Rogers Center for the Arts (North Andover, MA)

The Rogers Center for the Arts opened in the fall of 1999 on the campus of Merrimack College in North Andover, Massachusetts. It is not only an example of a quality multipurpose facility, but an example of a facility that was completely designed with acoustics in mind. This 600-seat theater is surrounded by concrete block walls and corridors to minimize the intrusion of background sounds from adjacent rooms. The mechanical room is also separated and isolated from the theater by the corridor and most of the HVAC ductwork is outside the main theater area (as is shown in Figure 6.13). Although the basic shape of the room is rectangular, the side walls are angled in sections to reflect sound into the audience area (see Figure 6.14). A removable orchestra shell (see Figures 6.15 and 6.16) is set up on the stage for appropriate concerts and a permanent set of wooden cloud-type reflectors hanging in front of the stage (see Figure 6.17) enforces sound coverage to the audience for all events. Figure 6.18 shows that these hanging reflectors are tilted at different angles with respect to each other to spread sound more evenly to the audience.

The reverberation time in the hall can be adjusted by drawing or opening six panels holding heavy curtains. These banks are in the rear and on each side wall of the facility, as shown in Figures 6.19 through 6.22. When the lowest reverbera-

Figure 6.13   HVAC ductwork and mechanical room separated from the main theater by a corridor and concrete block walls. *(Photograph by Joan McQuaid.)*

Figure 6.14   Rogers Center for the Arts, view from the stage toward the audience. *(Photograph by Joan McQuaid.)*

tion time is desired, as for a lecture or drama presentation, all curtains are exposed, and when the highest reverberation time is desired, as for a symphony concert, all curtains are retracted behind the walls.

The main sound system for the facility is hidden behind acoustically transparent fabric across the top and sides of the stage, as shown in Figure 6.17. An additional cluster of loudspeakers is hung midway above the audience, on a time delay from the main system, to enhance sound coverage to the audience seated toward the rear of the theater. Its black color allows this secondary cluster to be hidden from view, especially with the lights and black ceiling, rigging, and ductwork surrounding it.

By all accounts, this hall has effectively taken advantage of acoustical principles. This is reflected through the praises of both musicians and audiences. Merrimack College had no real facility for the performing arts before 1999, with its limited repertoire being performed in the school's gymnasium or church basement. It now has a world-class facility that is attracting top talent from all directions.

Figure 6.15 View of the stage with the top of the orchestra shell in place. *(Photograph by Joan McQuaid.)*

Figure 6.16 The sides of the orchestra shell in pieces behind the stage. *(Photograph by Joan McQuaid.)*

Figure 6.17 Wooden suspended reflector panels in front of the stage. Note that the main sound system is behind the cloth panels directly above the stage opening. *(Photograph by Joan McQuaid.)*

Figure 6.18 View of the stage from the audience, showing the tilted pattern of the wooden clouds in front of the stage. *(Photograph by Joan McQuaid.)*

Figure 6.19 Side wall panel without absorptive curtains. *(Photograph by Joan McQuaid.)*

Figure 6.20 Side wall without absorptive curtains. *(Photograph by Joan McQuaid.)*

Figure 6.21 Same wall as shown in Figure 6.20, with absorptive curtains. *(Photograph by Joan McQuaid.)*

Figure 6.22 Rear wall panel section with curtain exposed. *(Photograph by Joan McQuaid.)*

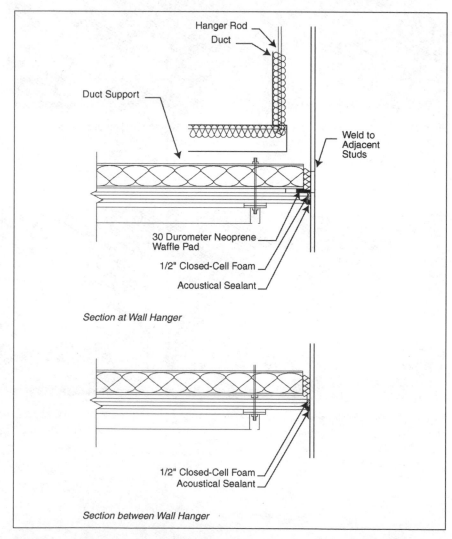

Figure 6.23 Isolation design for rooftop mechanical system noise in the Baltimore Convention Center.

Case Study: Mechanical System Noise Control— Baltimore Convention Center (Baltimore, MD)

In this case, a chilled-water plant is sharing the same space as the Baltimore Convention Center. Mechanical equipment associated with the plant is mounted on the roof of the convention center and the mechanical system noise can be clearly heard in some of the meeting rooms below. Since high levels of noise and vibration needed to be isolated and the building was already constructed, three to four layers of resiliently supported gypsum wallboard were recommended for a new ceiling, with at least a 3-in (8-cm) thickness of glass fiber batt insulation above them. The ceiling was to be resiliently suspended from the structure, using spring and neoprene isolators. These isolators should not contact the ductwork, and the perimeter of the multilayered ceiling should be isolated from the building structure using resilient materials, as shown in Figure 6.23. Flow velocities also needed to be kept to a minimum to minimize noise generation in the system.

On the roof of the building, the equipment needed to be isolated from the building structure by mounting the entire substation unit on neoprene pads. The entire surface of the pads needs to be loaded uniformly to ensure proper transfer of the load to the pads. As with all the recommendations in this book, this covers the general principles. It is critical that people experienced in this field specify the actual isolation materials necessary for these installations.

One key issue of this project was that there were many potential noise sources and it was difficult, if not impossible, to isolate the contribution of each. This is a common situation with noise control challenges, especially in constructed buildings. It is important to address each noise source separately, with the most obvious first, and to gradually eliminate noise sources as problems until the noise level is acceptable. This may take some time but can save building owners significant costs and reconstruction headaches.

CHAPTER 7

Sound Outdoors

There are several issues about sound travel and its control outdoors that deserve attention. When sounds generated by an outdoor facility are loud enough, they can be heard at great distances at times. The specific level of sound at distant listeners is highly dependent on the atmospheric conditions between the source and each listener. The control of sound outdoors is also much more difficult than indoors because of refraction and diffraction. The case studies in this chapter illustrate some limitations of outdoor sound travel and control that are helpful to know when designing projects involving outdoor areas.

Design Tools

Purpose	Design Considerations	Potential Problems	Solutions
Outdoor sound reduction	Minimize noise generation to surrounding community.	Noise is nuisance at sensitive properties.	Minimize noise at source; position barriers close to source and within 200 ft (61 m) of listener; enclose source if feasible.

Purpose	Design Considerations	Potential Problems	Solutions
Long-distance sound travel requirements	Develop sound system to project large distances.	Sound is too loud close to source and/or not loud enough at remote listeners.	Mount high-powered sound system at high elevation or seek alternate means of communication.*
Community noise issues from enclosed buildings	Minimize noise generation to surrounding community.	There is a nuisance or noise ordinance violation.	Employ effective noise control measures at the source and/or in the building design.

* When a warning system is required to cover a radius in terms of miles, atmospheric conditions can cause this system to be inaudible at some distant locations. When these types of distances are required for sound travel, it is often more reliable to use radio-based communications.

Case Study: Amphitheater—Chastain Park (Atlanta, GA)

The Chastain Park Amphitheater was originally an open-air theater in the park used for drama presentations. Gradually, musical performances began at the facility and the Atlanta Symphony also used the facility for concerts. At the point when pop concerts began playing at the facility, enough sound was escaping into the surrounding resi-

Figure 7.1 Chastain Park Amphitheater. *(Photograph by Robin Henson.)*

dential communities that complaints caused the owners to redesign the facility. One advantage in the facility's original design was that it was sunken into the ground, providing a natural barrier to the surrounding communities. Although this helped reduce the sound escaping from the facility, it was not providing enough sound reduction for the powerful sound systems used by touring rock bands.

Several schemes were suggested, including sound barriers, performance restrictions (in terms of allowable sound level limits), and a distributed loudspeaker system. The facility management decided on the option of a new sound system that focused sound more to the audience than outside the facility. Concerts at the original facility were equipped with towers of loudspeakers on the stage, pointed to the audience and played at high level. The new system consists of many loudspeakers along the sides of the theater, pointed into the audience and each played at a lower level than the original speakers.

Figures 7.1 and 7.2 show the layout of the facility.

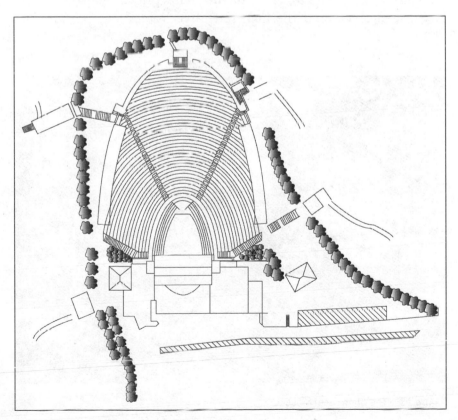

Figure 7.2 General layout of the amphitheater facility.

Case Study: Distance Amplification— Shah Alam Mosque (Selangor, Malaysia)

Shah Alam is a town recently built outside the capital of Malaysia, Kuala Lumpur, that serves as the capital of the Malaysian state of Selangor. The Shah Alam Mosque is the prayer center of the town, with a prayer hall covering an area of 75,000 ft² (7000 m²), a dome that is 250 ft (76 m) above the floor, and courtyards and verandas covering an area of 180,000 ft² (16,700 m²). The corners of the mosque are flanked by four 450-ft (137-m) minarets, as are shown in the picture of the facility in Figure 7.3.

The two most significant acoustical issues concerning this facility were speech intelligibility in the main prayer hall (given the large open space and the high reflective domed ceiling) and the

Figure 7.3   The Shah Alam Mosque.

desire for the call to prayer to be heard 3 mi (5 km) from the facility. The acoustics indoors was addressed by using many low-level, directional loudspeakers mounted on columns and by having appropriate electronic delays programmed. To preserve the aesthetics, the loudspeakers were built into the columns, as shown in Figure 7.4.

The issue with the call to prayer was more complicated. One of the minarets was equipped with high-powered loudspeaker systems at a 300-ft (91-m) elevation. This system included loudspeaker horns and compression drivers capable of sending audible sound at least 3 mi (5 km), considering atmospheric absorption of sound. One aspect that could not be considered in the sound system design, however, was refraction caused by temperature variations in the atmosphere. The typical temperature pattern in this area decreased with altitude at the call-to-prayer times, causing sound waves to bend up in the atmosphere and generate shadow zones on the ground. The result

Figure 7.4 Delayed loudspeakers mounted on columns in the main prayer hall.

was that the call to prayer was audible only to 2 mi (3 km) from the mosque.

Since we cannot control atmospheric conditions at this juncture in science, our recommendations were that existing sound systems in neighborhood mosques be tied into this system (with appropriate time delays). One other question to answer was why the sound level was basically uniform over the 2-mi (3-km) coverage area. People expected that the sound level would be loudest closest to the mosque; however, that was not the case because the loudspeakers were high in elevation and aimed away from the ground below. The loudspeaker horns were also chosen to focus the sound energy in a tight directional pattern.

Case Study: Community Noise Issues—
Harvard Faculty Club
(Cambridge, MA)

The campus of Harvard University, like that of most universities, consists of a dense concentration of buildings of all uses. This case study illustrates the control of noise from rooftop ventilating equipment so as not to bother nearby residents. In this case the local municipality was receiving complaints from a resident living across the street from the Harvard Faculty Club. These complaints were related to noise generated by the kitchen ventilation and exhaust system on the roof of the facility. The resident who complained lived in a multistory apartment building and her apartment overlooked the rooftop equipment. When local authorities came out to measure the noise levels generated by the equipment, their measurements showed that the local noise ordinance was being violated. The club then needed to reduce its generated noise levels to comply with the town's noise ordinance.

Several noise control measures were taken. The stack openings that had been facing the apartment building were replaced with vertical stacks. These stacks were also fitted with silencers, as are shown in Figure 7.5. As an additional measure, metal and wooden barriers were constructed on the sides of the roof facing the apart-

Figure 7.5 Acoustic silencers mounted on ventilation stacks on the roof of the Harvard Faculty Club. *(Photograph by Joan McQuaid.)*

Figure 7.6 Noise barriers on the roof of the Harvard Faculty Club, as viewed from across the street. *(Photograph by Joan McQuaid.)*

Figure 7.7 Noise barriers on the roof of the Harvard Faculty Club, as viewed from down the street. *(Photograph by Joan McQuaid.)*

Figure 7.8 Rooftop mechanical equipment mounted on resilient pads for vibration isolation. *(Photograph by Joan McQuaid.)*

ment building to break the line of sight between the apartments and the mechanical equipment. These are shown in Figures 7.6 and 7.7. Although the tops of the stacks can be seen over the barriers, their noise is being significantly reduced by the exhaust silencers.

Other noise control measures of note for this equipment are the spring and resilient mountings shown in Figures 7.8 and 7.9. These are effective in isolating vibrations generated by the equipment from the building below. Yet another effective noise control measure is the wrapping of ductwork with lagging material, as is shown in Figure 7.10. This material is

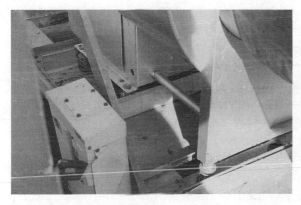

Figure 7.9 Rooftop mechanical equipment mounted on springs for vibration isolation. *(Photograph by Joan McQuaid.)*

Figure 7.10 Ductwork wrapped with lagging material to reduce sound transmission. *(Photograph by Joan McQuaid.)*

specifically designed to reduce noise radiated from this type of ductwork.

The combination of these measures reduced noise levels to below the local noise ordinance limits, which, in this case, required a reduction of only 5 dBA. If a higher level of noise reduction had been required, enclosing the equipment might have been necessary.

Case Study: Highway Noise Control—
Route 85 Noise Mitigation Study (Santa Clara County, CA)

Noise generated from vehicles traveling on highways affects more people than any other outdoor environmental noise source. New residential developments are being built within 100 ft (30 m) of interstate highways all over the United States. Home buyers are being told by developers that they will not hear the highway noise because barriers or earth berms will be placed between their houses and the roadways. This fallacy is resulting in many legal battles throughout the country.

This study was performed for the California Department of Transportation (Caltrans) to analyze options that could be used to reduce highway noise along a section of Route 85 near San Jose. Noise barriers were already in place in many areas; however, people were disenchanted with their acoustical performance (since the limits of barrier noise reduction effectiveness are usually less than 15 dBA). This study was chosen for inclusion in this book because it explores many noise reduction options, both practical and impractical. In this way, the practicality of each method is explored. Each proposed noise reduction method was rated for noise reduction effectiveness, engineering feasibility, maintenance, aesthetic impact, and cost. This case study should be instructive for anyone considering options for highway noise control.

Noise reduction methods were considered at the source, in the path between the source and listener, and at the listener. Source considerations included vehicular noise control, traffic controls, road pavement materials, and enclosed roadways.

Since the dominant noise sources for surface highway vehicles are the engine exhaust and tire-pavement interaction, the most effective vehicle noise control would involve quieting engines and tires. The cost of technology and legislative enforcement would be prohibitive for these measures. Traffic controls can be used to restrict speeds and types of vehicles on roadways (since trucks generate more noise than cars). In terms of speed restrictions, a decrease in speed of 25 mi/hr (40 km/hr) has been estimated by Caltrans to result in a decrease of 5 dBA (1-hr energy average) in traffic noise level. Noise level reductions with variations in vehicle mix have not been documented. These types of restrictions are not practical for a highway.

Road pavement materials have been shown to reduce average noise levels by up to 4 dBA; however, since these "quiet" pavements involve tining that could get clogged by dirt and salt, maintenance is key to their noise control effectiveness. Enclosing roadways would be quite effective for noise control but is usually prohibitive in terms of cost and engineering feasibility.

Noise control measures considered in the path between the traffic and the listener included berms, extensions of existing barriers, absorptive treatments for barriers, special barrier tops, landscaping, and active noise control. The greatest limitation of berms is the space that they take up. There is often not enough room between homes and highways for berms. When there is enough room, they provide an attractive option, but, just as with barriers, their noise reduction effectiveness is limited to 15 dBA. Berms do not provide a significant improvement over barriers. Therefore, replacing a barrier with a berm is not a practical option.

As is mentioned in earlier chapters, a barrier must break the line of sight between a source and a listener to provide any sound reduction to the listener. After the line of sight is broken, the noise reduction effectiveness of a barrier increases (from 5 dBA) at a rate of roughly 3 dBA for each additional 3 ft (1 m) of barrier height, up to a limit of roughly 15 dBA (for the homes closest to the barrier). Because of variations in topography, barrier heights greater than 25 ft may be required just to break the line of sight to listeners. Barriers taller than 25 ft can introduce structural issues that make their construction impractical.

Absorptive treatments for faces of barriers are effective under only two conditions: (1) when barriers are on both sides of a highway and are parallel to each other, or (2) when barriers are on one side of a highway but the side not protected by a barrier has residences on it. In each of these two cases, reflections can cause additional sound in areas where it is not welcome. In the case of the parallel barriers, sound waves reflect between the barriers and degrade the performance of the barriers on each side of the road. In the case of the single barrier on the opposite side of the road, sound reflects off the barrier directly into the residential area. Absorptive treatments would reduce, if not eliminate, these effects. If neither of these cases exists, though, absorptive treatment would be of no acoustic benefit.

Special barrier tops have been designed with different shapes (such as a T, a Y, or a cylinder) that may provide a few dBA of noise reduction in comparison to flat barriers; however, more field-testing of these products is necessary to document their true effectiveness.

Landscaping provides minimal noise reduction, unless room is available for at least 100 ft (30 m) of dense evergreen forest. Active noise control, at this juncture in science, is completely impractical for moving outdoor sources.

Figure 7.11 A brick highway noise barrier in southern California.

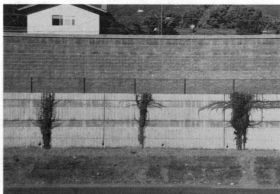

Figure 7.12 Second floor of residential building unprotected by noise barrier because of a clear line of sight to the highway.

Figure 7.13 A concrete highway noise barrier in New York. Note that the wall must block the line of sight to truck exhaust stacks to effectively reduce truck noise.

Figure 7.14 A wooden highway noise barrier. Note that the second floors of the houses are not protected by these barriers. They also do not have a long service life due to warping (causing air gaps) and disintegration.

Control of highway noise at the listener is limited to insulating houses. The most impractical aspect of home insulation is the fact that it deals only with indoor sound levels and does nothing to reduce sound levels outdoors.

The main recommendations from this study were to reduce speeds by 10 mi/hr (16 km/hr), to resurface the road with a quieter pavement, and to increase the heights of barriers less than 12 ft (3.7 m) tall to up to 20 ft (6 m) tall.

Figures 7.11 and 7.12 show the existing barriers in the study area. To emphasize a few points, Figures 7.13 and 7.14 have been included. Figure 7.13 is an example of a typical concrete noise barrier in the northeastern United States, emphasizing that the height of the barrier must take into account the height of truck exhaust stacks. These noise sources may be up to 12 ft (3.7 m) above the highway surface. Figure 7.14 shows a typical wooden noise barrier, revealing several potential problems. Of prime importance are the low height of the barrier and the short useful life of the material.

Technical Addendum

On the "Goodness" of Acoustics

BY REIN PIRN

Every serious musician and every serious listener wants good acoustics. Sometimes they have it, but often they do not. What is good acoustics? What makes one hall or one situation sound better than another? The question has haunted us for decades, yet the answer remains clouded—not because it is unknown, but because it is ill understood.

Above all, acoustical quality is a perception. An experienced ear will separate the good from the bad, which the eye does not. Therein lies a common danger. There are more than a few musicians and music lovers whose perception is influenced, if not dominated, by what they see rather than what they hear. A "warm" appearance, as conveyed by the materials and colors in a hall, may override deficiencies that the eye cannot see. Or the musical importance of the occasion may be such as to overshadow all else. In both cases, good sound is expected and, in the opinion of many, achieved.

Perceived excellence in any one hall on any one occasion may be but a fleeting experience. If it recurs, and if it is publicized, only then does a hall acquire a reputation for excellence. That, of course, is the dream of every musician, hall manager, and architect—to play in, to operate, or to have designed a good hall. They then believe that the acoustics is good, but is it really?

The scientist may take a different view. A good scientist, specifically one conversant in room acoustics, will look for (and hear) nuances that go undetected by others. He or she is analytical and precise. The language used is numbers—reverberation times, energy ratios, infinitesimal time patterns where a second is eternity—and not vague terms such as "rich," "brilliant," or "mellow." Yet those are the descriptors that matter, regardless of the numbers.

There is validity in both camps. Acoustical quality cannot be had if the hall's physical parameters are not within a range of objectively predictable and measurable values. Nor is it achieved if the results are judged to be less than good. Favorable numbers alone are not enough, just as a good impression (often abetted by the visual scene) may be largely an illusion.

Let us review a few facts. Most musicians have heard of reverberation time. They may even know its value in some "good" hall and therefore assume that longer or shorter reverberation times are suspect. Not necessarily so. Today there is ample evidence that reverberation time is more the result of other—more important—parameters being on target than an indicator of acoustical quality. As for the other parameters, popular knowledge of what they are and what they mean is scant.

Or take the case of materials. There is hardly a musician who does not believe that wood is good. (After all, many fine instruments are made of wood!) It must, they argue, enrich the sound with warmth that is so evident to the eye. It may, but—perish the thought—so does any other equally sound-reflective material. The reverence for wood is so strong and so widespread that despite the acoustical dangers (loss of bass due to resonance) it must rank high among the materials of which concert halls are made.

What about the scientific view? Fair enough, reverberation time merits some consideration, but there is much more: the so-called clarity index, the lateral energy fraction, and steady-state sound levels, to name a few. Each is an independent part of a complex whole called acoustical quality, and each can be controlled by design. Clarity—a seemingly desirable property—can be overdone. Lateral sound—that is, strong reflections from the listener's left and right—apparently cannot be overdone; they give rise to a "spatial impression" that only recently has emerged as a most desirable quality. And then, there is the question of loudness, which is so dependent on the hall's design—especially its size. The limits of acceptability are still not well defined, but music can be too loud for comfort or, conversely, too feeble to arouse the audience.

The problem—and a major one—is that even if everything were perfect, the conditions cannot be perfect for the entire musical repertoire. It is usual to judge major halls by the way they respond to traditional works (e.g., the symphonies of the romantic period), as performed by a conventional orchestra of some 80 to 90 players. But what about the music of other periods, and ensembles of substantially different size or composition? Each has its own acoustically ideal setting which no single hall can duplicate. Compromises must be made and accepted. Expectations must be tempered with the realities that preclude perfection for all occasions.

Finally, but very important, let us not forget the artists. The sound they hear on stage is quite different from that heard by the audience in the hall. It is louder, less well blended, and not subject to the same criteria that apply to the hall at large. All this is self-evident, if one considers these simple facts: musicians are in the midst of the sound they make; they are at very different distances from each other; they make the music and thus appreciate qualities that help them play better. Their world is the stage and its acoustics, and not so much the whole hall.

What are the musicians' wants? For one, they must hear their own sound, or else intonation may suffer. But they also must hear the other sections of an orchestra, or else synchrony may suffer. They need quick reflections of their own as well as their fellow musicians' sound. The far reaches of a responsive hall also return sound to the stage, which many players appreciate, but not if their immediate environment—the stage itself—is unresponsive. Delayed reflections alone can be quite disconcerting.

It is indeed remarkable how much more is known about concert hall acoustics today than a mere decade or two ago. The qualities that contribute to goodness have been and are being redefined. However, tradition also lives on. Perceptions and beliefs so dear to the hearts of many (for example, the magic of wood) may never die. Except for the blind, whose judgments are based solely on what they hear, appearance still matters. Rare are the cases that appease both the ear and the eye, without fooling either.

Sound Performance in Public Spaces

BY CARL J. ROSENBERG, AIA

Three separate groups of kids are practicing basketball in the gym and the din is deafening—how is it that they seem to be making more noise than a Saturn rocket? There's a close finish to the relay race at the swimming pool, but you can't distinguish the announcements because the noise is so garbled. The clatter in the bank lobby makes waiting in line uncomfortable and you can't understand the teller when it finally is your turn. A finely presented elegant (and expensive) meal is ruined by the distracting noise in the restaurant.

These public spaces share similar acoustic problems: sound in the room is too loud; it is harsh and raucous; it appears to come from all directions at once, adding to the sense of cacophony and disorientation; and intelligibility, articulation, or understanding of speech is poor. These types of spaces do not require the finely tuned acoustics of a concert hall, but they can be designed for acoustic comfort. It is possible to investigate the acoustic performance of "nonperformance" public spaces, present a model for the way sound responds in such spaces, and then apply treatments to control acoustics.

The model can be applied to similar spaces that share the following attributes: they are not performance or presentation spaces; intelligibility of live (unamplified) sound is not critical (there is no orchestra or dramatic presentation); the sound sources are spread out or distributed throughout the space; and the dimensions of the room are relatively even—that is, the space is not unduly long, wide, or high. Given these attributes, sound (which travels at 1120 ft/sec) will rapidly fill the space with a diffuse, evenly distributed sound field. For example, in a room 60 ft long, the sound will bounce from one end to the other and back 10 times each second. In a smaller room, the diffusion and buildup of sound occur even more quickly.

Outdoors, where there are no reflecting surfaces, sound radiates and expands spherically, like the ripples on a pond or like a balloon being blown up. Sound radiating outward can be diagramed as a series of arrows or rays projecting in all directions from a source. Sound is measured on a logarithmic decibel scale—with every doubling of the distance from the source, the sound level decreases by 6 dB. In the model of sound as a ray, the reflection of that ray off a hard surface follows the simple rule that the angle of reflection equals the angle of incidence. In other words, sound reflections behave like light. Occasionally, when a signal comes back to us after at least one-tenth of a second, we hear discrete reflections as echoes, even though the sound has become quieter—by 6 dB every time the distance from the source is doubled.

Inside a room, the walls bounce the sound waves (or rays) back and forth, creating the diffuse sound field already noted. However, we can still model this sound field as if it were composed of discrete reflections or images. Each sound ray has an energy level that depends on how far it travels as it bounces around (that is, the distance from the receiver to the sound image) and on the absorptive properties of the surfaces the ray encounters. We hear multiple reflections as reverberation or the smooth decay of sound in a room, but, for purposes of this model, reverberation is made up of many individual reflections.

The designer can lower the level of built-up sound energy in a room by lengthening the distance a sound wave has to travel to reach a receiver's ears—that is, making the room larger—or by changing the physical characteristics of the surfaces that reflect the sound wave—that is, adding absorptive materials.

Understanding sound absorption is a key to using this simple model. As a sound ray hits the surface of any material— wall, floor, ceiling, table, or computer—that surface absorbs some sound. This is a basic property of physics caused by the interaction between a pressure wave (sound in air) and a boundary layer of a surface that has an impedance different from that of air. The efficiency with which a material absorbs sound is measured in a laboratory as a coefficient of absorption, designated α. The coefficient of absorption is expressed as a value between zero and one, representing the percentage of the sound absorbed.

The efficiency with which a material absorbs sound varies with the frequency of the sound. Most materials are much better at absorbing sound at high frequencies than at low frequencies. This is related to the size of the sound waves and the thickness of absorptive materials.

For most architectural applications where human speech is the predominant source of noise, it is sufficient to design a space using an average value of coefficient of absorption. This is called the Noise Reduction Coefficient (NRC), and it is an average of the coefficients of absorption at 250, 500, 1000, and 2000 Hz. Because the NRC value is simply an average, it is rounded off to the nearest 0.05. NRC values are a useful tool to give the designer a rough idea of how well a material absorbs sound in the frequency range of speech.

Higher NRC values indicate better efficiency at absorbing sound. Materials with high NRC values usually are soft, porous, and fuzzy, because open pores allow the sound pressure fluctuations in the air to release their energy as friction to the material. Good choices are glass fiber, acoustic tile, shredded wood fiber form board, spray-on cellulose, curtains, cloths, some carpets, and so forth. Because they tend to be rather delicate and expensive, good sound-absorptive materials often are premanufactured into components that also offer some protection against wear and tear.

At the other end of the spectrum, materials that are efficient at reflecting sound (and therefore have low NRC values) usually are hard and smooth, such as wood, gypsum board, and glass.

For the model of a diffuse sound field in an enclosed space, we are concerned with the average NRC of all the surfaces together because, by definition, sound is diffusely reflected within the enclosure and hits all the surfaces quickly. Experience and extensive theoretical analyses have shown that the average NRC of typical architectural enclosures with all "hard" surfaces—such as wood, metal decks, gypsum walls, and furniture—is about 0.1, meaning that 10 percent of the sound is absorbed by the boundary surfaces. Some extremely reverberant, live spaces might have average NRCs for all surfaces that are less than this. For example, a ceramic tile bathhouse might have an average NRC for all surfaces that is only 0.05.

Consider our noisy gym, with hypothetical dimensions 50 ft long by 80 ft wide and 22 ft high. The total area of all exposed surfaces is 13,720 ft^2, calculated as follows: 4000-ft^2 floor (50 × 80); 4000-ft^2 ceiling (50 × 80); and 5720 ft^2 of wall surface (two walls of 50 × 22, two of 80 × 22).

Assume that the surface area has an average NRC of 0.1, which includes the wood floor, metal ceiling, bleachers, lights, and all other surfaces. To increase the average NRC to about 0.25, we could treat all the surfaces—floor, ceiling, walls, and bleachers—with a material of this value, such as carpet. Acoustics might be superior, but we would not have a workable gym.

We can get the same acoustic result by treating about one-third of the surface area with a moderately absorptive finish of NRC 0.7. A similar value results if we treat only one-quarter of the surface with a superefficient material of NRC 0.99.

The numbers work because the sound field is diffuse for this type of space. We are not concerned with a specific reverberation time, although this could be calculated and is related to the average NRC. Nor are we concerned with a specific noise level, although this too could be calculated for a given noise source. Rather, we are using a model of a diffuse sound field to implement reasonable noise control. The goal is simply to obtain an average NRC for all surfaces equal to 0.25 or greater.

Because the sound field is diffuse and sound bounces off all surfaces everywhere so quickly that we can catch it anywhere, this model does not relate the absorption to any particular location. But common sense suggests either spreading the absorptive material around the room evenly, so that it captures some sound from all reflections, or placing it close to the source or receiver, so that it reduces energy mainly in the first or last reflection. In that light, the ceiling of a public space often is the best place for sound-absorptive material. It covers the whole room, it comprises usually about one-third of the total surface area, it is out of the way and relatively safe from abuse (important for expensive and fragile materials), and it will not be covered with tables and chairs or interrupted by windows and doors.

In gymnasiums, if adequate absorptive treatment is incorporated as part of the ceiling system early in the design process, lots of potential problems can be avoided. If the ceiling of the gym is also the roof, an acoustic metal deck is an excellent choice. Sound-absorptive glass fiber batts are placed between the webs of the deck and the webs are perforated so that the insulation is exposed to sound. The NRC values of these systems are often above 0.8, so, if the ceiling is one-third the surface area, an overall average NRC of 0.24 is easily achieved.

Other options for gym ceilings include a suspended acoustic tile ceiling (with hold-down clips so that bouncing balls do not dislodge the tiles), shredded wood fiber form board, or sound-absorptive baffles suspended vertically from the joists. In any case, such treatment of the ceiling often will be all that is needed to reduce the noise to reasonable levels, making the gym a suitable place for instruction and spectator viewing.

If the dimensions of the gym are such that the ceiling does not provide enough absorption area even when covered with a very efficient material, it may be desirable to add materials with high NRC values on the upper side walls. These might be glass fiber panels covered with a perforated vinyl or regular building insulation protected by shredded wood fiber form board. Durable materials that can withstand the abuse of volleyball impact are available for this use.

Swimming pool areas, of necessity, have floor and water surfaces that are very reflective of sound, and their lower wall areas also must be of solid materials such as ceramic tile. But here, too, an average NRC of 0.25 can be achieved if other surfaces, especially the ceiling, are treated with very efficient sound-absorptive materials, such as metal deck systems. Other suspended ceiling systems are specially designed for high-humidity and corrosive environments or employ suspended elements such as perforated metal panels with glass fiber wrapped in thin plastic laid above the panels.

Television studios generally have dimensions that meet the criteria for a diffuse sound field. Typically, the studio space must be quite dead, so the average NRC for all the surfaces must be at least 0.25. Floor surfaces may have to be vinyl to be smooth enough for cameras to be wheeled around without vibration. To achieve an average NRC of 0.25 or greater for all surfaces, the design must incorporate an absorptive ceiling. Curtains around the perimeter walls also will help raise the average NRC if they are bunched up (100 percent fullness, or 2 ft of curtain per foot of wall), spaced 4 to 6 in away from the wall, and made of a heavy material, say, 18 to 25 oz/yd^2.

Some degree of noise control for a studio space can be achieved by increasing the volume of the space. The distance that reflections would have to travel would be increased if the roof were raised and the walls placed farther apart; hence, the decibel level of each reflection would be lower, by 6 dB if the distance of the sound-ray path is doubled. The main problem with this approach is the great expense of increasing volume just for acoustic results. Another problem is that increasing volume often multiplies the noise sources within the space. For example, during the 1987 World Series, the built-up noise levels in the enclosed Metrodome in Minneapolis were extremely high, even though the volume of the space is huge, with the reflective surfaces being far apart and the ceiling having a very efficient sound-absorptive design.

Atriums and lobbies also have high volumes but usually do not hold crowds cheering for a World Series team. The same acoustic model can be applied here for noise control. The high volumes help reduce noise buildup, even where there are many hard surfaces. Therefore, the average NRC can be reduced to 0.2 and still yield an acoustically comfortable space. Less absorptive material may be needed to give a quieter feel to such a space.

Restaurants and cafeterias are notorious for being noisy, but if the ceiling can be adequately covered with sound-absorptive material to hold the average NRC to 0.25 or so, most problems will disappear. Acoustic tile is one way to do this, although more elegant finishes may be desired, such as cloth-covered glass fiber panels floating at the ceiling. The walls will seldom provide enough area for absorption to make any difference and even carpet on the floor will not be adequate. Additionally, the floor material is covered by tables, which, even with the most elegant tablecloths, do not have a high enough NRC to help.

Keep in mind that for nonperformance public spaces the sound-absorptive materials must be spread out over all the surfaces, and the average NRC of all the surfaces is what determines whether there is adequate noise control. This model, plus some common sense and some care in selecting materials and checking on their acoustic properties, can make for happy clients and successful projects.

Gothic Sound for the Neo-Gothic Chapel of Duke University

BY ROBERT B. NEWMAN AND JAMES G. FERGUSON, JR.

As a preface to discussing the Duke University Chapel project, let us begin with 1916 and a letter from a partner of Cram, Goodhue, and Ferguson—one of the leading early twentieth-century architectural firms.

BERTRAM GROSVENOR GOODHUE ARCHITECT
West Forty-Seventh Street New York, N.Y.

Rafael Guastavino, Esq.   New York
949 Broadway, New York   May 5, 1916

My dear Mr. Guastavino:

On Easter Sunday I attended the dedicatory service at the First Congregational Church at Montclair. To the best of my knowledge and belief no such acoustical result has ever been achieved before except possibly by accident.

Whether form, proportion and dimensions can from now on be absolutely disregarded I do not know; but it is certain that by the substitution of your acoustic tile for the more usual plaster and stone interior walls the acoustics of any church, music hall or auditorium may be made almost, if not quite perfect.

To you and Dr. Sabine all credit is due and it is difficult to express my satisfaction with the result of the years of patient effort spent by you both in the perfecting of this wholly new material.

Please accept my thanks and congratulations, and believe me,

Very faithfully yours,
(signed Bertram Goodhue)[1]

Only with the added information that the "acoustic tiles" referred to by Goodhue had an NRC of 0.45 to 0.50, do we become aware of the true acoustical ambience of the New Jersey church. It was an ambience created over and over again—in the Duke University Chapel; the Princeton University Chapel; the National Cathedral, Washington, DC; the University of Chicago Chapel; Grace Cathedral, San Francisco; St. Thomas Church, Riverside Church, and the Cathedral of St. John the Divine, New York City. The list could go on, but it is unnecessary.

Goodhue's letter not only sketches the outline of this ambience, but also gathers together two of those men most responsible for its creation—Rafael Guastavino and Wallace C. Sabine. There is no need to discuss Sabine here. The work of the Guastavinos, however, may be somewhat obscure and merits a brief digression.

It is a simple historical fact that many of the great secular and religious spaces built in this country between the 1880s and the 1940s would not exist if it were not for the work of the Guastavinos.[2] While our primary concern is acoustics, it is clear that we would not have a Duke University Chapel project to discuss without Guastavino structural techniques.

Before coming to the United States in 1881, Rafael Guastavino had distinguished himself by perfecting and exten-
sively using a Mediterranean vaulting form known as Catalan
vaulting (for Catalonia, the Spanish province). In contrast to
the traditional masonry vault, which used heavy stones and
elaborate wooden centering and falsework, the timbrel tile
vault is constructed of two or three layers of tile set in a fast-
drying Portland cement mortar. In spite of their light weight,
these vaults were found to have structural properties that
were superior to those constructed of stone or brick. Evi-
dently this superiority existed because the tiles were laid in
such a way as to cover the breaks of the previous course of
tile and mortar. In addition, Guastavino felt that the mortar's
rapid setting as well as its freedom from centering and falsework contributed to the strength of the vaulting.[3] Since the
mortar did set so quickly, it was possible for workmen to use
a previous day's completed tile course as a platform upon
which to work for the next day. This attribute led to such
widely publicized tours de force as photographs of workmen
setting in the tile courses in the crossing vault of St. John the
Divine with only the thin tile course laid the day before
between them and the stone floor 150 ft below.

The combination of light weight and fire resistance (the
Guastavino Company was at one time called the Guastavino
Fireproof Construction Company)[4] made possible construction that would have been prohibitive in steel and concrete at
the time. Additionally, as Collins notes in the following quo-
tation, the combined strength of the tile and mortar appeared
to surpass that of any other known structural material:

> The Metropolitan Museum vaults presented considerable
> difficulty in their removal. On several visits to the demoli-
> tion, the present author failed to find a single whole tile in
> the rubble; the aggregate had proven to be so homogeneous
> and rigid that pneumatic drills were being used with confi-
> dence by workers standing on unsupported remnants of the
> vault that jutted out as much as eight feet.[5]

Small wonder that Guastavino vaults were used as supports for the Henry Hudson Parkway as well as neo-Byzantine, neo-Romanesque, and neo-Gothic celestial visions.

While Rafael Guastavino laid all of the technological groundwork for the vaulting construction, it was the collaboration between his son, Rafael II, and Sabine that produced the material we are discussing here—Akoustolith.

The immediate predecessor to Akoustolith was a material named Rumford tile. Patented in 1914, this tile was produced from a clay/peat mixture, which, when fired, attained porosity because of the burned-out peat particles. Used in many locations, this material's most significant application is in St. Thomas Church, New York City.[6] In 1916, Akoustolith tile was patented. In this case, the porosity was produced by pumice particles bonded together with a very dry mix of white Portland cement and set under pressure. As a final development, Akoustolith plaster was patented in the 1920s.[7]

One might be tempted to ask why thought was even given to building such dramatic religious spaces so clearly modeled after European precedents if it was intended that the "cathedral sound" be muffled at the outset. The answer, of course, lies in the importance given to preaching in the United States as compared to Europe. One does not typically hear 30-minute sermons at Bourges and Chartres. Perhaps the medieval architects were more clever than we imagine! By the same token, the primitive state of public address systems in the 1920s and 1930s demanded the suppression of reverberation if the clergy were to be heard. Hence Akoustolith—patent #1,197,956. Hence, incidentally, the building of pipe organs having wind pressures occasionally reaching 50 in (water gauge) and stops with treble ranges having three or four unison pipes per note.

Designed by Horace Trumbauer of Philadelphia and constructed from 1930 to 1932, the Duke University Chapel is of a neo-Gothic design and is similar to the Princeton University Chapel. The building is cruciform in plan with the following interior dimensions: combined nave and chancel length of 264 ft; widths of 30 ft in the chancel, 54 ft in the nave, and 112 ft at the transepts. The attached Memorial Chapel is approximately 54 ft long and 26 ft wide, and the octagonal narthex is approximately 30 ft wide. The crown of the crossing vault is 75 ft from the floor, the nave vaulting centerline is 73 ft from the floor, and the Memorial Chapel vaulting centerline is 50 ft from the floor. The volume of the space is approximately 1 million ft^3.

Structurally, the building is one of the purer examples of revival-style architecture in that it makes use of steel only as a means of transferring thrust to exterior buttresses. The columnar supports are of limestone, as are the vaulting ribs, arches, and window mullions and quoins. From floor level to a height of about 8 ft, the nave sidewalls are finished in limestone, from which point they are finished in a material composed of a 1-in Akoustolith tile bonded to a 1-in slab of concrete. This material covers the remainder of the walls in the nave, narthex, and Memorial Chapel. It also serves as the soffit in the vaulting throughout the building. (For some reason not disclosed by Trumbauer, the chancel walls were finished in limestone—perhaps at the behest of the organ builder.) A rough determination yields the estimate that there are over 25,000 ft^2 of Akoustolith in the room. This is almost half the total surface area of the space.

In 1969, Duke University contracted with D. A. Flentrop of Holland to build a new organ at the nave-narthex junction and it agreed to improve the acoustics according to the recommendations of Bolt Beranek & Newman, Inc. (BBN) and to the satisfaction of Flentrop.

A preliminary examination of the chapel interior revealed that the only reasonable means of achieving the maximum potential reverberation time was to seal the porous surface of the Akoustolith. At this point the only known treatment for sealing Akoustolith tile to date had been a somewhat limited project undertaken at the Riverside Church, New York

City, in consultation with BBN. Thus, while the nature of the treatment for Duke may have been clear, the prescribed formula was not. Also, the chapel administration was told at the outset that improvement of the acoustics for music would degrade hearing conditions for speech, thus necessitating the replacement of the existing multiple-source speech reinforcement system by a system using distributed directional loudspeakers and time delays.

After consulting a member of the civil engineering faculty at MIT in October 1969, BBN concluded that an acrylic emulsion (Rhoplex) manufactured by Rohm and Haas would be the most appropriate sealant. After considerable experimentation, impedance tube testing had shown that two undiluted coats of the emulsion were required to seal the surface and produce the very low absorption coefficient of the limestone surface of the rest of the chapel interior. Aesthetically, however, this solution left something to be desired, in that there was a glossy finish on the treated surfaces. Subsequent investigation by the Kyanize Paint Company resulted in a satisfactory second coat of sealant having a matte finish which matched the limestone. As a practical matter it was not feasible to extrapolate spreading rates from the 4-in tile sample used in the impedance tube tests, so the painting contractor was instructed that the treated Akoustolith ". . . must have no remaining pin holes or voids of any sort, and must really be filled."[8]

Before the walls were treated, the authors felt that both archival interest and the necessity to measure acoustical change required assessment of the existing chapel acoustics. Accordingly, reverberation time experiments using balloon bursts and organ chords in the chancel and narthex areas were conducted in September 1970 (see Figure T.1).

In March 1971, a second reverberation time measurement was taken (see Figure T.1). At this point the nave had been entirely treated according to the initial specification with two coats of sealer, but the chancel, narthex, and

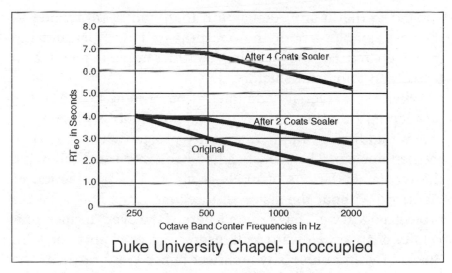

Figure T.1 Measured RT_{60} before and after sealant treatments in the Duke University Chapel.

Memorial Chapel were yet to be painted, and the crossing area was obstructed by the painters' scaffolding. This measurement was taken to confirm Flentrop's observation that the apparent acoustical change was far less than he expected and certainly not satisfactory for the organ he was to build. In fact, the data showed that the absorption coefficient of the Akoustolith at 500 Hz had been reduced from approximately 0.5 to 0.25. This, however, was a far cry from the coefficient of 0.02 for the limestone.

It was clear that the all too common difference between laboratory and field procedures was at fault. While the painters had attempted to achieve a complete visual seal, discussion with the foreman revealed that achieving this seal by using two coats of paint was a practical impossibility for painters lying on their backs on a slightly unsteady scaffold 75 ft off the stone floor in 90°+ temperature. In addition to dealing with less than optimal working conditions, the painters were also concerned about the paint "running" down the walls and creating visual problems. All of these considerations led inadvertently to less paint being applied to the in situ Akoustolith than had been applied to the test samples—both the one prepared by BBN and the one prepared by the painting contractor, which was also tested and found satisfactory.

When the university administration became aware of the developing situation, they were understandably reluctant to commit additional funds to a costly project whose outcome seemed unclear. The acoustics had been improved following the recommendations of BBN—if not to the satisfaction of Flentrop. Without the vision of three men, the Duke Chapel project might well have ended here. However, former university chaplain Howard Wilkinson, former chancellor John Blackburn, and university architect James Ward resolved the quandary of how to proceed.

Using impeccable experimental technique, Wilkinson devised and executed a number of trial applications of the sealant in order to come up with a practical solution for the apparent impasse. By simulating the interim condition of the painted tile, he was able to demonstrate that two additional coats—a heavy one of base sealer and a light one of the flat sealer—would provide a visual seal.[9] Furthermore, his studies provided a practical directive that the painters could follow as well as a criterion that would enable the university to monitor the quality of the second treatment.

Although Wilkinson's experimental evidence seemed conclusive, the university was appropriately cautious and asked that an additional set of acoustical measurements be made on the test patches done by the painting contractor on the chapel walls. Accordingly, a field impedance tube test was performed in January 1973 by I. L. Ver of BBN on samples treated with additional coats of sealer (see Figure T.2). As the figure indicates, the experimental data show that the two additional coats of paint actually produced an absorption coefficient slightly lower than that of

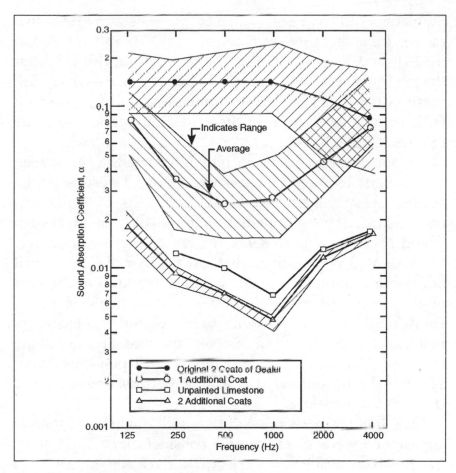

Figure T.2 Field measurements of the treated Akoustolith.

the limestone, but this difference is within the range of measurement error and therefore probably not significant. On the basis of these confirmatory acoustical measurements, a contract was let for the retreatment of the entire Akoustolith interior according to Wilkinson's findings— one heavy coat of base sealant and one light coat of flat sealant.

When the second treatment was concluded, a final reverberation time measurement was taken by BBN in September 1973 (see Figure T.1). There was no doubt that the desired dramatic results had been achieved. Reverberation time now ranged from slightly over 5 sec at 5000 Hz to 6.75 sec at 500 Hz and 8 sec at 40 Hz. The test data confirmed what the choir members and organists already knew—Duke University Chapel had been transformed into a liturgical space without equal in the United States. Gothic sound had come to a neo-Gothic building.

As noted earlier, the concomitant degrading of speech intelligibility required the replacement of the existing public address system. Even before the final coats of sealer had been added, it was apparent that the old sound system would never be able to serve the more reverberant space. We suggested two systems that could serve the space well and confine amplified speech sound to the level of the ears of the congregation. One, like the system installed in St. Paul's Cathedral in London, used column loudspeakers mounted on the limestone piers at the crossing and along the nave. The other called for about 500 pew-back loudspeakers throughout the seating area. Both systems involved the use of time delays.

In examining both possibilities, performance, cost, and appearance were given careful consideration. Ultimately, the pew-back system was rejected for reasons of cost and performance. It was clear, for example, that the pew-back system could not deal adequately with the large area of movable chairs in the crossing. There was, however, equal concern about the appearance of the column loudspeakers in a space that was essentially the way Trumbauer left it in 1930. This concern was put to rest by the university architect, who had full size mock-ups of the speakers installed in several nave locations to assess public reaction. The reaction was minimal and the column loudspeaker system is now a matter of history.

Once the column arrangement was selected, we examined the demands that would be placed on the sound system.

Clergy and music staff alike were asked to describe the ways in which they used (and might use) the space. Thus, microphone inputs were placed at all significant liturgical locations in the building, as well as in the new organ gallery. In response to a specific request by music staff, a distributed system using pew-back loudspeakers was designed for the chancel to enable choir members to hear speech from the lectern and pulpit. Considerable attention was also devoted to electronic media needs, inasmuch as the chapel services are broadcast weekly and musical recording in conjunction with the new organ was contemplated.

In response to the specifications provided by Duke, a sound system was designed that used 18 column loudspeakers in addition to approximately 50 pew-back units installed in the chancel.

The column loudspeakers were custom-made enclosures 8 in × 8 in × 50 in tall, each of which contained nine 3-in loudspeakers. Fiberglass wedges were interposed between the speakers and grille cloth to ensure greater uniformity of distribution pattern at all frequencies. These loudspeakers were mounted on each of the structural piers at both sides of the nave as well as on the piers near the pulpit and lectern to serve the crossing and transepts. All the enclosures were tilted downward toward the congregation at about 15°.

The loudspeakers serving the six bays of the nave were driven by four sets of amplifiers with four time delays ranging from 20 ms to 110 ms. Loudspeakers in the Memorial Chapel, crossing, narthex, and chancel were driven without delays. The systems were equalized with eight plug-in ⅓-octave bandwidth equalizers installed by the contractor. Although final confirmatory data on performance were not taken, subjective impressions and informal communication suggest that the system is highly reliable and performs with excellent realism and quality.

References

1. George R. Collins, "The Transfer of Thin Masonry Vaulting from Spain to America," *Journal of the Society of Architectural Historians* 27 (October 1968), p. 195.

2. Ibid., p. 176.

3. Ibid., p. 177.

4. Ibid., p. 180.

5. Ibid., p. 183.

6. Ibid., p. 195.

7. Ibid.

8. Newman to Ward, December 12, 1969.

9. Wilkinson to Blackburn, April 28, 1972.

Blocking the Bombardment of Noise

BY CARL J. ROSENBERG, AIA

In our modern, industrialized society, we constantly are bombarded by a wide range of noise from the environment, such as from airplanes, cars, trucks, buses, cooling towers, lawn mowers, air conditioners, and so on. In a school, high levels of noise from nearby roads will discourage learning. In houses near an airport, the deafening sound of aircraft awakens people, disrupts telephone conversations, and interferes with television viewing. At a hospital, the noise from a helicopter landing pad can be stressful to patients. At offices, intrusive noise can reduce productivity.

To wage the battle against these pervasive sounds, we need a measure of their magnitude. Our measurement tools describe the spectrum (noise signature of pitch and frequency), the level (loudness), and the duration (time patterns) of environmental noise.

Spectrum

Each noise source has its own unique sonic characteristics, which let us distinguish a truck from a motorcycle, or one type

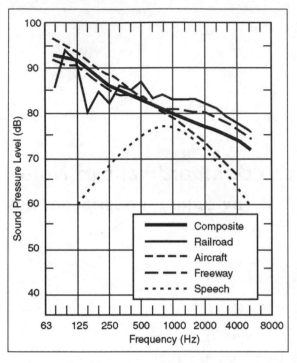

Figure T.3 Average frequency spectra for environmental noise sources.

of automobile engine from another. Nevertheless, recent studies have shown enough similarity among all these spectra to suggest a standard spectrum that represents in a reasonable manner the average of freeway, aircraft, and railroad noise sources. This average spectrum allows us to model the noise in the environment for a number of sound reduction studies (see Figure T.3). The salient aspect of this average spectrum is that most of the energy is low frequency, around 250 Hz and below.

Level

The loudness of a sound is measured in decibels (dB) for each frequency range. Because the spectra of environmental noise sources are so similar, we can simplify our analysis, combining all the frequencies using the commonly accepted A-weighted filter of a sound-level meter. This filter adjusts and combines the sound in the same manner as the human ear, which discriminates against low frequencies. This means that, for equal levels of acoustic energy, we are less sensitive to low, rumbling sounds than to high-pitched sounds. Without this filter, the low-frequency sound of diesel engines (which is similar to the spectrum in Figure T.3) would be deafening. A-weighted noise levels in the environment can range from 30 dBA (very quiet) to over 100 dBA (painful).

Duration

The noise from modes of transportation (airplanes, cars, trucks, trains, helicopters, subways) fluctuates over time. The fluctuations may occur over just a minute or two (as a truck goes by), or over an hour (when there is pile driving or demolition at a construction site), or over a full day (noise levels from highway traffic vary as road traffic patterns change), or over a year (varying levels around an airport due to seasonal wind patterns).

To resolve this dilemma, acoustic engineers have developed a measure called the *equivalent noise level,* or L_{eq}, which represents with a single number, for any given time period, the equivalent or average noise energy from fluctuating sources. The decibel scale is a logarithmic scale, so the energy average is not an arithmetic average. For example, imagine a 20-min interval when for half the time the noise level is 70 dBA and for the other half the sound is quieter, at 50 dBA. The L_{eq}, or energy average, for the 20 min is 67 dBA, not 60 dBA.

L_{eq} values typically are measured for an hour and can range from L_{eq} 40 (perhaps during the quiet nighttime hour in a rural setting) to 90 dBA or more (perhaps near a construction site during excavation).

For workplace locations, the daytime L_{eq} values are an accurate description of how loud the noise is. For residential areas, it is necessary to account for the fact that, because of normal sleep patterns, we are more sensitive to noise late in the evening and early in the morning. This is done by the day-night equivalent noise level, or L_{dn}, which averages 24 hourly L_{eq} values over a full day, but first penalizes or increases the nighttime values (occurring between 10 P.M. and 7 A.M.) by 10 dBA. The L_{eq} and L_{dn} values are the most commonly used metrics for comparing and evaluating environmental noise. L_{dn} values range from 45 (a very quiet enclave) to 80 dBA or more (close to an airport).

The actual L_{eq} or L_{dn} noise exposure level at a site can be measured with proper acoustical gear, or it can be estimated with adequate data on the noise source patterns, such as traffic counts, train operations, or whatever. Also, airports are required to document the noise exposure to the community with Ldn contours updated for current and proposed future runway utilization. The L_{eq} or L_{dn} for every site level does exist and can be determined.

Noise Exposure Criteria

Studies in the early 1970s by the Environmental Protection Agency suggested that a reasonable threshold of annoyance for environmental noise in residential areas for most people in a normal population is around an L_{dn} of 65 dBA. When environmental noise exceeds an L_{dn} of 65 dBA, many people find it objectionable. This value has been promulgated by the U.S. Department of Housing and Urban Development (HUD) and many state housing agencies to be a limit for "acceptable" conditions.

This acceptability is based on the assumption that typical residential constructions provide about 20 dBA of sound attenuation and that a suitable interior exposure should not exceed an L_{dn} of 45 dBA. When exterior environmental noise levels exceed an L_{dn} of 65 dBA, HUD and others require additional sound reduction measures to protect the interior environment.

Of course, everyone seems to respond to sound differently, depending on the level of the sound, one's association with it, one's lifestyle, and the message of the sound. At one site near a railroad track, planners were perplexed at the low incidence of complaints in spite of high noise levels; then they found that most of the residents used to work on the railroad and found the noise comforting.

There are other environmental noise metrics and criteria one may encounter, such as speech interference levels (SIL, the arithmetic average of background levels in the speech frequency range), the noise level exceeded 10 percent of the time (L_{10}), the community noise equivalent level (CNEL, the L_{dn} with a 5-dBA penalty added between 7 P.M. and 10 P.M.), and others, which are similar to the basic L_{dn} concept but different in particulars.

The 65-dBA L_{dn} criterion relates only to environmental noise sources. Local noise codes address concerns about steady-state sources, such as industrial activity or your neighbor's air conditioner; these community noise codes may be more stringent, perhaps allowing only an instantaneous level of 50 dBA or so at the property line.

Also, our sense of annoyance may be quite unrelated to actual noise levels. For example, in one town, the community basketball court mistakenly was located near some houses; the activity was annoying because of the content of the sound (the thump, thump of the basketball and associated expletives), and the court had to be removed. Research has shown that the single most annoying noise source in residential communities is barking dogs.

The best noise control treatment is to quiet the source. For example, there are codes, albeit poorly enforced, for noise from diesel exhausts. Cooling towers can be run at reduced speed during the night to quiet the noise; with multicell towers, two fans at half speed are quieter than one fan at full speed, and they deliver the same amount of cooling. Construction noise can be limited to certain hours of operation. FAA regulations are phasing in quieter aircraft engines. Highways often are bordered by berms or barriers to contain the noise. (Trees and plantings that hide noise sources have absolutely no acoustic benefit, but they can reduce our sense of annoyance by removing the offending source from view.) When we are limited in the amount of noise control we can introduce at the source, we look to buildings to protect us from noise.

Soundproofing Strategies

Most of us accept reasonable levels of environmental noise inside our residences and places of work. However, in environments where the noise is unduly loud—that is, where exterior levels typically are 65 dBA L_{dn} and above, or where people are particularly sensitive—improvements must be made to the noise reduction of the building envelope. As with all acoustical problems, this means treating the weakest path first. In addition to tackling the problem at its weakest point, there is a hierarchy of lines of defense that suggests strategies.

The Sound Transmission Class (STC) rating is the accepted nomenclature for the ability of materials or construction systems to block sound. The advantage of this single-number terminology is its universal acceptance by architects and engineers and the availability of performance data (although many exterior wall and roof constructions have not been adequately tested). The disadvantage is that the STC rating is designed to rate the noise reduction performance of materials for typical office or speech sources, not for environmental noise (as is the standard spectrum described here and shown in Figure T.3), which has much more low-frequency content than speech. We can still use STC ratings, but keep in mind that they usually overstate the insulating properties of a construction system by 5 to 10 dBA for environmental noise. That is, a construction with a rating of STC 30 will reduce environmental noise by only about 20 dBA.

The primary path through which sound attacks a building is where there are no barriers at all, that is, the cracks and gaps and leaks around penetrations. These openings have a value of STC 0. Old windows that rattle in their frames or ungasketed doors are significant sound leaks. Sealing these leaks is the same primary concern as when the goal is preventing heat loss during the winter. It does not make sense to insulate the walls until you close the door. When a typical building is sealed in this manner, we usually achieve 20 to 25 dBA of noise reduction.

The next-weakest sound path acoustically is usually the windows. Single-paned glass has a noise reduction value of about STC 25 to 30 (that is, 15 to 20 dBA of environmental noise reduction). Insulating glass is not much better, because the narrow air space between layers of glass, less than an inch, effectively couples the two layers of glass together, acting like a connector to form a single window unit.

In all fairness, glass really is an exceptional sound barrier material, as it is very dense. For equivalent thickness, it is better at blocking sound than concrete. However, we do not usually see 6-in-thick slabs of glass.

As a second line of defense, the way to provide more protection against higher environmental noise levels is to improve sound isolation of a window system by using two completely separate layers of glass spaced at least 2 in apart and preferably 4 in apart. This large air space allows the layers to act more independently and improves the STC ratings by 10 to 15 points. More important, it improves the noise reduction performance at low frequencies, to 250 Hz and below. In residential applications, this can be done with well-fitting, tight storm windows either inside or outside. Nevertheless, the large air space is necessary for sound isolation. There also are special sound-isolating double windows, which have two separate panels connected in one frame, with a proper thermal break (see Figure T.4).

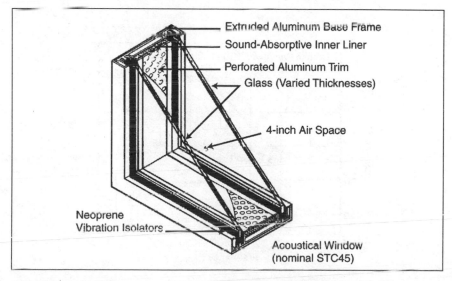

Extruded Aluminum Base Frame
Sound-Absorptive Inner Liner
Perforated Aluminum Trim
Glass (Varied Thicknesses)
4-inch Air Space
Neoprene Vibration Isolators
Acoustical Window (nominal STC45)

Figure T.4 Cross section of a properly designed acoustical window.

Another window treatment is to use acoustic laminated glass, similar to the safety glass on car windshields. The lamination allows the two parts of the window sandwich to move independently and to be damped in their movement. This has a benefit of about 5 to 10 STC points compared with a similar thickness of regular glass.

Laminated glass comes in thicknesses of ¼, ½, and ¾ in (STC 36, 39, and 42, respectively) and can be combined into regular insulating glass configurations or into a double-window system (see Figure T.5).

In one recent project for an office building near a railroad, the preferred sound-isolating system was a double window with a 4-in air space. However, the visual impact of the deep mullion on the facade was unacceptable to the historic preservation review commission. Instead, the overall window thickness was reduced to 1¾ in, using two layers of laminated glass and a smaller air space, and it still achieved a 35-dBA reduction of train noise (a rating of STC 46).

Other penetrations must be considered, too. Residential doors should be solid core, well gasketed, or weather-stripped. If possible, there should be a vestibule; if not, a

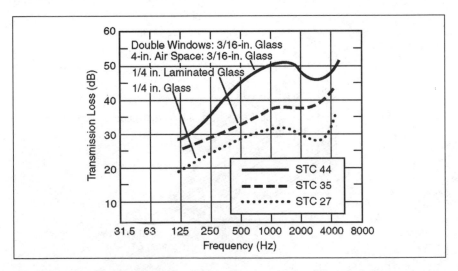

Figure T.5   Comparison of sound reduction effectiveness for different window types.

good storm door. Air conditioners are a problem because, when the vent is open to bring in fresh air, the STC of that path is 0, and when the vent is closed, there is only a single layer of sheet metal as a barrier; such air conditioners or vents may have to be removed or the ventilation system changed to a central system so that noise control techniques can be added properly. These techniques can provide 30 to 35 dBA of environmental noise protection, a 10-dB improvement that will make intrusive sound levels seem half as loud.

In most houses, the noise reduction performance of good windows and storms will approach that of the rest of the envelope. There may not be any remaining single weakest path. What can be done, then, as a third line of defense for additional improvement? In this case, all systems must be upgraded together, most likely with double construction systems that provide a dead air space between inside and outside. This will entail double windows (not just storm windows), double walls (separate inner studs or inner skins on resilient channels), and separate ceiling structures. Attic insulation becomes important. Vestibules are now the mandatory way to treat doors. The inner room becomes a separate room in the building.

This concept of a room-within-a-room is just as suitable for other, more severe, noise isolation problems as well. For performance environments, the criteria for excluding intrusive noise are extremely severe. It may be necessary to develop a full isolation system for walls, windows, doors, roof, and even the floor. St. Peter's Church in the Citicorp Complex in New York City sits entirely on neoprene isolator pads so that the noise and vibrations from the subway below are not audible.

There are two ways the conditions within a room can affect annoyance from intrusive noise. One is that absorptive finishes in a room reduce the sound somewhat, but they affect only the sound that has already come through the walls or windows. A heavily furnished bedroom will sound slightly quieter than an unfurnished bare room; unfortunately, we have found that the amount of sound absorption in typical living spaces does not vary greatly once the room is occupied. The maximum difference might be no more than 3 to 5 dB, and seldom do we see such extremes.

The second manner in which the interior environment can affect annoyance is the degree to which other sounds are present. Part of our annoyance from environmental noise is that it is recognizable and draws our attention. In one office building, where an executive was distracted by the truck traffic outside, the additional window we first proposed was going to be too expensive and awkward to install. Instead, we added a sound-masking system that introduced a noise of its own, louder than the highway noise, but not so loud as to be distracting. The constancy of the masking sound covered the fluctuation of the traffic sound. When you are in a hotel near an airport or interstate highway, a fan may be more soothing than the roar of trucks or the whine of a jet engine. This same approach is even more successful outside, as shown in the vest-pocket parks in New York City, where fountains help cover street noise.

Quieting Multifamily Dwellings

BY CARL J. ROSENBERG, AIA

*O*ver the past few years, there has been a surprising increase in complaints and concern about sound isolation and privacy in multifamily dwellings. As acoustical consultants, we have seen a great number of lawsuits brought by condominium associations against developers and architects for inadequate sound isolation performance, and lawsuits against realtors or sellers for misrepresentation of so-called soundproofed units.

This trend is probably exacerbated by the growing number of condominium conversions of older buildings. Apartment dwellers can accept a certain amount of intrusion and lack of privacy; after all, the building is not theirs. But the owners of what used to be apartments—now called condominiums—expect and demand greater privacy; they can't blame the landlord anymore. We also see higher expectations from older persons who are moving from suburban single-family homes to multifamily buildings in the city, or to life-care facilities. This demand for acoustical privacy puts pressure on developers, who are pushed to reduce costs and minimize expenses, even for so-called luxury housing. Additionally, it is becoming common practice for building codes to require that the architect demonstrate a specified level of acoustical performance for building components in certain occupancy types, including multifamily dwellings.

The most prevalent acoustical problems in multifamily dwellings are airborne sound isolation between adjacent units, both horizontally and vertically; impact noise from the occupants upstairs; and plumbing noise.

Avoiding these problems begins with an understanding of how to evaluate sound transmission from one building space to another. The most commonly used term for evaluating a building construction for its performance at blocking sound is the Sound Transmission Class, or STC, rating. This rating is based on the decibel scale, and values can range from near STC 0 (no blocking of any sound) to as high as STC 80 (virtually no sound transfer). In typical construction we commonly see values between STC 30 and STC 60. At the lower end of this range, you can hear normal conversation through a wall without much trouble. At the higher end, even a high-powered shouting match would be a mere rumble coming from the other side of the wall. However, such blocking performance is rare.

Any wall type or building material can be tested in a laboratory to determine its STC value. However, when that same construction system or material is installed in the field, you can expect a rating that may be 5 to 10 dB lower. One reason is that the laboratory test sample was built by laboratory technicians, not a general carpenter. In the field, details like caulking may be omitted, there may be back-to-back outlets, and there are invariably other flanking paths (such as a common floor) that transmit sound from one side of the wall to the other by way of the surrounding structure, without the sound going through the wall itself.

The STC is a rating for a single construction element, tested in a laboratory. Actual field performance is more accurately described by the Noise Isolation Class, or NIC, which is a measure of the overall reduction of noise from a source room to a receiver room, irrespective of how the noise traveled.

The STC and NIC rating systems, which are based on the same procedure and are thus in a way equivalent, were designed to measure how effectively a sound-isolating construction blocks human speech. They are not accurate evaluation or measurement tools for sound from a mechanical room next to a bedroom, or for isolation of aircraft noise, or for other special cases where the source of potential annoyance is not speech. However, for the usual airborne noise sources in multifamily housing, the STC and NIC systems are a good guide for comparing different constructions, keeping in mind that there may be adjustments between laboratory and field-test data.

Wall Construction and Vibration

Sound waves striking one side of a wall will cause the structure to vibrate, and the vibrating wall will radiate the sound wave to the other side. The STC value will increase (that is, less sound will get through the wall) if the vibration is decreased, either by a heavier wall (which better resists the sound wave) or by decoupling of the surfaces (which breaks the structural continuity).

In practical terms, what then makes a good party wall between adjacent dwelling units? The most important ingredients are:

- Adequate mass
- Separation of the skins of the wall
- Absorption in the cavity

Assume, to start, standard stud construction for the party wall, with a single layer of gypsum board on each side. This might yield a field performance of NIC 35. The following questions and answers pertain to increasing the NIC value.

Does it matter whether the stud is wood or metal? Probably not much, although light-gauge metal studs seem to be less rigid and stiff than wood studs, and this helps decouple the wall.

Does it matter how many layers of gypsum board are on the stud? Yes, more are better because of the extra mass. Doubling the layers of gypsum board will come close to doubling the weight of the wall, which will improve sound blocking by 5 or 6 dB (STC points), which is quite significant. Furthermore, the extra layers offer an opportunity to stagger the joints, thus reducing the chance for cracks and gaps that leak sound.

Should there be insulation in the cavity? Yes, but with a qualification. If the studs are wood, the insulation merely damps the gypsum board, which then vibrates less. Isolation improves by 3 to 5 dB. If the studs are metal—that is, if the two sides of the wall are already somewhat decoupled—the improvement can be greater, generally 5 to 8 dB.

Is plaster better than gypsum board? Only the weight really matters. Gypsum board weighs approximately 4 lb/ft^2 per inch of thickness; plaster weighs 5 to 9 lb/ft^2, depending on the aggregate used.

Is it worth using ⅝-in gypsum board instead of ½-in or ⅜-in? Again, only weight really matters, and only if you can increase the weight by about a factor of two. In other words, changing from ½-in to ⅝-in does not do much good, but changing from one layer to two makes a significant difference.

Is caulking necessary? Absolutely, because hidden behind every baseboard or molding is a potential unseen opening with a relative STC value of 0, waiting to let sound pass through the wall.

What about back-to-back outlets? These are another potential leak through an otherwise decent wall. And phone jacks and antenna outlets can be even worse, because they may not have even the benefit of back-boxes, so they are literally holes in the wall. Therefore, it is helpful to offset the openings so that they occur in different stud cavities.

Does it matter what type of insulation is in the wall? Not much. Glass or mineral fiber (with or without paper lining, with a density of about 3 lb/ft^3), rock wool, or vermiculite all contribute about the same benefit of soaking up sound in the cavity and damping the gypsum board facing. Thicker batts are generally better than thinner ones, provided they fit loosely and are not crammed into the cavity.

Keep in mind that the best single-stud wall will be only marginal for sound isolation as a party wall, because it has that one integral stud that will transmit sound energy from one rigidly attached skin of the wall to the other. We have found the field performance of a single-stud wall—even with two layers of gypsum board on each side and insulation in the cavity—to be in the range of STC 40 to STC 45, which is far below the performance that most guides and laboratory data would have you expect.

For better performance, there should be more mass or more separation. More mass could mean concrete or concrete block. In this regard, dense, heavy concrete offers better sound isolation than lightweight concrete of equal thickness. More separation entails finding some way of decoupling one side of the wall from the other.

RESILIENT CHANNELS

Resilient channels are metal furring strips with holes punched in the middle web to make the channel less rigid, thus helping to decouple the faces of the wall. In a laboratory test, especially with wood studs, they may improve sound blocking by 7 to 10 dB. But to achieve this full acoustical benefit in the field, the resilient channels must be installed carefully, and this is often awkward and difficult. Proper installation requires no "short circuits," or places where the gypsum board might be in contact with the studs.

Cabinets hung on the wall and baseboards nailed through to the studs will reduce the isolation integrity of the channels. Seldom do we see satisfactory wall installations of resilient channels.

STAGGERED STUDS

In wood construction, it is possible to align 2 × 4 wood studs on a 2 × 6 plate so that alternate studs support opposite sides of the wall. (This is not possible with steel studs because the inner flange of the stud has no part of the plate to be fastened against.) Staggering the studs improves the sound isolation performance, but there is still a common plate and header.

DOUBLE STUDS

A separate row of floor-to-ceiling studs for each side of the wall can provide 10- to 15-dB improvement over the performance of a single-stud wall. Contractors find this an easy wall type to install, and it has the potential for excellent sound isolation performance.

This lesson was well learned by a developer who had sold a new town house unit to a lawyer, assuring him that the unit was soundproof with respect to the neighbors. The party wall was single-stud construction with insulation in the cavity. When the lawyer was not happy with the noise he heard from his neighbor, the developer first tried to remedy the problem by adding gypsum board to one side of the wall. Then he added furring strips and resilient channels and more gypsum board to the other side, with insulation in the new cavity. The lawyer persisted in his complaints, but there was no more room for additional construction to be added to the wall, and its performance was still below NIC 42. Finally, the developer had to buy back the unit at a loss, much to his dismay.

This developer sought advice on his next project before building it. He carefully erected party walls of double-stud construction. He even had each side of the wall carried on separate footings and kept all framing separate from unit to unit. The resulting performance was above NIC 58, which is exceptionally good. The developer then could proudly demonstrate to prospective buyers the acoustical excellence of the model units.

FIRE-STOPS AND GUSSETPLATES

Obviously, any rigid contact between one stud and the other will reduce the acoustical benefit of the double-wall system, and therefore must be detailed to maintain separation between separate walls.

Floors and Ceilings: Vertical Transfer

The same concerns of mass and separation apply equally to floor/ceiling constructions. If the flooring above and the gypsum board below are rigidly attached to a single set of joists, chances are good that sound will transfer easily from one floor to the other.

Won't insulation in the cavity solve all problems? With the rigid ties through the joists, the vibrations will bypass the insulation, and the improvement will be only 3 to 5 dB.

How can I add more mass? A concrete topping or slab construction or a poured gypsum underlayment will improve the sound isolation performance if the mass is sufficient. A thin, lightweight topping (such as one used as a leveler) probably will not do much good by itself.

RESILIENT CHANNELS

We find that resilient channels work much better for ceilings than for walls because the channel can be correctly and consistently loaded by the ceiling all the time, and it is harder to short-circuit the installation. Resilient channels are by far the most useful and practical way to achieve separation of the layers in a floor/ceiling system. Remember, the perimeter of the ceiling should be caulked with a nonhardening acoustic caulk so the ceiling can still move independently of the walls. Also, you must avoid recessed lights—they reduce the integrity of the ceiling and create acoustic holes in the otherwise solid barrier.

SOUND-DEADENING BOARD

So-called sound-deadening board usually refers to a ½-in mat of fiberboard sandwiched between the subfloor and the finished floor, which is supposed to help decouple the floor from the joist. This will not improve performance much because the sound-deadening board does not offer any resilience (it is not like a spring) and because the floor on top of the sound-deadening board is usually nailed through to the joists.

IMPACT NOISE

This is the sound of people walking around upstairs, or slamming doors, or creating any other impulsive source where the energy is induced directly into the structure. It is difficult to measure impact noise quantitatively because standard test procedures (such as the ASTM tapping machine test to measure Impact Isolation Class, or IIC) do not really duplicate the kind of energy generated by a typical source, such as a person walking. Also, impact noise may have varying components, such as the hard, high-frequency click-click of heels on quarry tile and the low-frequency thud or thump of footfalls on wood-frame construction.

For the high-frequency noise problem, the most effective solution, short of requiring the people upstairs to walk barefoot, is carpet, because it reduces noise at the source. Even with carpet, however, low-frequency thumping noises can be transmitted, especially in wood-frame dwellings. The only way to solve this problem even partially is to build a resilient ceiling as previously described, or use a heavier mass such as a concrete topping in the upper unit.

This heavier mass for the upper part of the floor will improve sound isolation and lessen impact noise transfer, and the mass can be even more effective if it is isolated or resiliently supported from the structure. In extreme cases this is done with resilient mounts for a so-called floating floor, but other products may be appropriate for residential conditions. A sound-reducing matting, which is a compression-resistant three-dimensional nylon pad, may be helpful when used with ceramic tile or a concrete topping, and with resilient channels for the ceiling below.

We have seen many old warehouses and wharves renovated in the Northeast. Architects look at the old plank floors and say, "Let's refinish them and leave them exposed." Then they look at the large beams and the underside of the plank below and say, "Let's clean these up and leave them intact." But invariably, to reduce impact noise, one or the other must be covered. In some cases the floor gets a topping of concrete with carpet and pad, with the edges carefully caulked (especially to fill in the gaps when the concrete shrinks). In other cases the ceiling is covered with resilient furring and gypsum board, with insulation in the cavity. For best results, do both.

Other People's Plumbing Noise

Plumbing noise (invariably from someone else's plumbing) is the bane of existence for most occupants of multifamily housing. It is not airborne noise that creates the problem—if you made a tape recording of the noise from a water fixture or shower and played it through a stereo in the neighbor's bathroom, at the same volume as the original, the noise probably would not intrude into the next unit. The problem is the vibrational energy of water in pipes that are in contact with the structure. This energy causes the studs and wallboard to radiate that energy as noise. The problem is exacerbated by PVC drains, which offer less resistance to the transfer of energy than do older, cast-iron pipes.

In any case, the vibrational energy must be decoupled from the structure. Back-to-back bathrooms should have completely separate framing, such as a double wall, so that one unit's piping does not contact its neighbor's. Similar double walls should be used wherever a chase wall adjoins a bedroom. Piping can be separated from the framing by wrapping with insulation and using oversized clamps. Chases should be heavily treated with insulation, perhaps 6 in thick if stud spacing allows.

In one luxury condominium project, the designer took great care to separate units with poured concrete slabs, double-stud party walls, and so forth. But in one place, where a duplex unit nestled under another unit, a bathroom abutted a stairwell and the wall was single-stud construction with a showerhead firmly attached to that stud. The shower noise made the stairwell sound like a waterfall. Remounting the showerhead solved the problem.

Background Sound, Rural and Urban

Studies conducted by the U.S. Department of Housing and Urban Development (HUD) in the 1960s highlighted the relationship of background sound levels to the degree of privacy that people could expect from neighboring units. It was found that a given wall construction in a rural environment, with very low background sound levels, would be less acceptable than the same wall construction in an urban setting, where background sound levels are higher. HUD therefore proposed that criteria for constructions between dwelling units be adjusted to account for the background sound level, with more stringent standards applied to rural locations.

Although this is still a good idea, it is better established in theory than in practice. Most codes do not yet recognize the contribution of background sound in covering up or masking sounds from a neighbor. Most builders discover this factor the hard way. Some perfectly good constructions, with high STC ratings, can be judged unacceptable if the background sound is so low that it provides no masking. A builder should be extra cautious with constructions in quiet areas—not only do they lack masking sound, but they are precisely the areas that attract buyers and occupants who equate quiet with quality. All is well until the neighbor moves in.

At a fine resort development, we encountered a modest floor/ceiling construction: open wood joists, wood floor above, gypsum board below. Complaints began as soon as one unit above another was occupied. The developer added insulation to the cavity; that helped a little but not enough. Then the developer totally rebuilt the ceilings with resilient channels, which raised the rating to near NIC 50. But occupants still are plagued with airborne and impact noise transmission. The units have electric heat, no sound of any air-handling equipment, and no nearby vehicular traffic. The background sound level is quieter than in some of the world's best concert halls, and as a result every footfall and pin drop can be heard. Such a quiet background requires higher than normal sound isolation or some masking sound producer, such as fans or air conditioners.

Planning and Layout

The worst acoustical problems, which are also perhaps the easiest to avoid, arise from poor layout and insensitive planning. Wherever possible, units should be stacked with like uses above each other, not just because it is neater but because it reduces acoustical problems and makes a better environment. After construction, there is no easy remedy for these conditions:

- A bathroom of one unit over the neighbor's living room, with the drains framed out below the living room ceiling

- A kitchen with an elegant quarry tile floor (a marketable amenity) over a bedroom

- HVAC equipment for one unit dropped into a soffit over the neighboring unit's family room

Common sense would have avoided these problems. Good sound isolation starts with good planning and layout. Next comes sensible design of wall and floor systems, using an understanding of the guidelines of mass, separation, and absorption. Also, keep in mind the background sound factor. Details, especially those involving resilient separation, must be executed carefully, and diligent supervision in the field is essential.

Soaking Up Sound: Properties of Materials That Absorb Sound

BY CARL J. ROSENBERG, AIA

In any room in which a sound wave is generated—any sound wave—the expanding fronts of the sound wave soon collide with materials on the surfaces of the room. Here, as a basic property of that material, three things will happen to the sound wave: it will be partially absorbed, partially reflected, and partially transmitted. This is a basic consequence of the material—any and all materials—and the interaction with the energy of a sound wave.

How Sound Is Absorbed

Sound absorption occurs when sound energy is changed into heat. This heat is not enough to fry an egg (although some loud bands seem to try to generate this much energy), but still, heat is generated. Three more commonly identified devices for turning sound energy into heat are porous materials, resonant-panel absorbers, and resonant cavity absorbers.

POROSITY (FLOW RESISTANCE)

Technically, when extensive surface areas composed of a porous semirigid framework with small openings, cracks, and crevices are impacted by the oscillating particles of a sound wave, flow resistance creates friction between the oscillating air molecules and the sound wave, and this friction generates heat. In common terminology, soft, fuzzy, porous materials can be great at absorbing sound. Glass fiber, mineral wool, open-cell foam, a thick cashmere sweater, and a plush carpet are examples of materials that employ this mechanism to varying extent.

However, the way these materials are mounted affects how they absorb sound energy. The flow resistance of these fuzzy materials is most effective at changing the sound wave into heat at the points of the sound wave propagation where the velocity is highest and where the pressure wave is least; that is, at one-quarter wavelength from a sound-reflective surface. Hence, the absorptivity of a material extends to lower and lower frequencies (that is, for longer and longer wavelengths) when it is spaced away from a hard surface or when it has added thickness (see Figure T.6). Glass fiber that is 6 in thick is a better absorber of low frequencies than glass fiber that is 1 in thick, although for high frequencies the absorption performance might be the same. In fact, to absorb low frequencies, you can take an inch of glass fiber, space it 6 in from the wall with an open cavity behind, and it will be pretty much as absorptive as if you used 6 in of glass fiber. The major benefit of the flow resistance offered by the absorbing material (in this case, the inch of glass fiber) occurs at the one-quarter-wavelength distance from the wall; hence, the spacing is as important as the material. In a similar manner, the absorption coefficients of acoustical ceiling tile vary depending on the spacing from the structure above. Suspended applications are necessary to absorb low frequencies. As yet another example, consider the effectiveness of felt cloth. If it were stretched tight on a pool table, the felt would never be considered as an effective material to absorb sound. But if it is held away from a wall and bunched up so that its multiple folds provide greater flow resistance, felt cloth can be a very effective sound-absorbing material.

These porous materials can be made less effective at sound absorption if facings are used. Most soft, porous, fuzzy sound-absorptive materials require a protective facing to be suitable for use in architectural applications. The percentage of open area of this surface determines how much the facing degrades the sound-absorptive performance, especially at high frequencies (see Figure T.7). In most cases, open, sound-transparent facings (such as grille cloth and other open-weave fabrics, wire mesh, and hardware cloth) have little or no effect on the sound-absorptive properties of a material. However, some facing materials begin to reflect high-frequency sounds (perhaps at 2000 Hz and higher), especially those frequencies for which the solid elements of the facing are of significant size relative to the length of the sound wave. This effect can occur with some types of perforated metal, wood slats, and expanded metal. Even if these facings are backed by efficient glass fiber, the solid areas will negatively affect the high-frequency absorption, which can be detrimental to particular applications, such as open-plan offices where high-frequency sound absorption is critical for improving the speech privacy between offices.

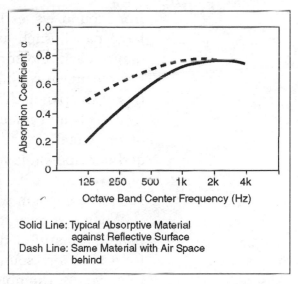

Solid Line: Typical Absorptive Material against Reflective Surface
Dash Line: Same Material with Air Space behind

Figure T.6 Added low-frequency absorption afforded by an air space between absorptive material and a surface.

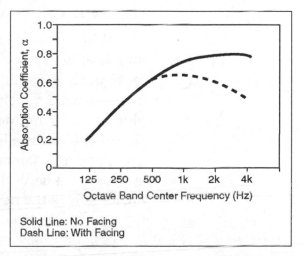

Solid Line: No Facing
Dash Line: With Facing

Figure T.7 Degradation of absorptive performance at high frequencies by facings.

RESONANT-PANEL ABSORBERS

Even materials that are not soft, porous, and fuzzy can transfer energy from sound waves into heat by their flexural movement. A large panel, such as a sheet of glass, plywood, or plaster, moves back and forth when a sound wave hits it, and this movement, in conjunction with the natural damping properties of the panel, dissipates the resulting vibration energy as heat. This movement typically occurs at a natural frequency of the panel, usually a low frequency below, say, 200 Hz, and is responsible for the characteristic "thunk" sound you hear when you bang gently against a panel construction, such as a gypsum wallboard.

When it is easy for the panel to move (for example, if the panel is unstiffened, plain, and lighter weight), it is easier for it to absorb sound. When it is more difficult for the panel to move (for example, if the panel is stiffer, thicker, and well braced) or when the cavity behind the panel is damped with absorption, it is more difficult for it to absorb sound. The more "thunk" you hear when you pound a wall, the more low-frequency sound can be absorbed. The fundamental resonance frequency of a panel absorber is affected by the stiffness of the panel, the depth of the cavity, and the absorptive material, if any, in the cavity.

RESONANT CAVITY ABSORBERS

The third mechanism by which sound energy is converted into heat is by viscous losses caused by friction with the movement of an air column at the neck of an enclosed volume—that is, a Helmholtz resonator. The movement of this plug of air at the neck of an enclosure is controlled by the size of the enclosed volume, the depth and diameter of the neck, and any damping material in the resonator. The typical system has a natural frequency, such as you hear when blowing air across the neck of an old-fashioned soda bottle. Resonant cavity absorbers are rather specialized sound-absorbing devices, but they are embodied in many premanufactured products, such as slit-faced, hollow concrete blocks, or are incorporated in air spaces behind perforated facings.

Data

It is possible to measure the sound-absorptive properties for all materials; this is not black magic. Standardized procedures, such as ASTM Test Method 423, have been used for years so that the same material tested in any approved laboratory will give about the same absorption coefficient results. For this test method, one would place at least 72 ft² (6.7 m²) of a given material in a large reverberation test chamber, measure the reverberation time before and after the material is placed in the chamber, determine the reduction in sound energy (related to shorter reverberation time) brought on by the introduction of the test material, and calculate the sound absorbed per square foot of material; this is the α for the given material. This procedure measures the random-incidence sound absorption because the sound field in a reverberant test chamber is extremely diffuse. Sometimes, because a sample laid on the floor has edges that increase exposed surface area and provide an increased area where diffraction can occur, or because the sound in the room is less diffuse when the sample is there, or because the product being tested has greater surface area than the area of the lab floor that it covers, the measurement procedure can create anomalies in the data such that coefficients seem to be greater than 1.00 or more than 100 percent of the incident sound per square foot in a given frequency band seems to be absorbed. Scrupulous reporting of such data will adjust the values to be less than unity so as not to mislead consumers. However, manufacturers often report absorption coefficients greater than 1.00 even though these values are caused solely by the measurement methodology. For further discussion, see "On the Reason for Random Incidence Sound Absorption Coefficients Being Reported as Greater Than Unity" by Angelo J. Campanella, presented at the Acoustical Society of America meeting, June 1994, in Cambridge, Massachusetts.

If you do not have a standard reverberation test chamber for sound absorption tests, you can still determine the absorption coefficients of a material by using a small sample at the end of an impedance tube. By measuring with a microphone probe the shape of a standing wave in a tube that has the sample at one end, we can deduce the impedance and normal-incidence absorption coefficient of the material. However, this measurement method evaluates normal-incidence sound absorption only and does not replicate the random-incidence conditions that more closely represent most room acoustics applications. The relationship between random-incidence and normal-incidence absorption coefficients is not always straightforward and will vary as a function of frequency.

To further complicate the issue, there is an increasing need for useful data on absorption coefficients for materials at specific specular reflection angles—for example, the ability of an acoustic ceiling tile to absorb sound in an open-plan office where speech impacts the ceiling at grazing incidence. Measurement standards for coefficients of absorption under these conditions are being developed.

Therefore, in reviewing data, one must be aware of the conditions under which the material will be used and the appropriate test values that should be applied. In any case, data are available or can be obtained for any material. Manufacturers should substantiate their product claims with valid, meaningful test data. Also, any test results should clearly state the conditions under which the material was tested, the ASTM standard, the size of the sample, the mounting conditions, the facing material, and so forth. Armed with such data, designers can compare the absorptive properties of materials in a meaningful way.

Other Conditions

Although the absorption coefficient tells us an important property of the acoustical material, in architectural acoustics applications, the total amount of absorption in a room is even more important. This factor depends on the materials and on how much of any material there is.

The unit for sound absorption is the sabin, A, and is determined by the product of the coefficient of absorption, α, of a material and its total surface area, S ($A = S\alpha$). In English units, S is measured in square feet; in SI units, S is measured in square meters. Absorption units then are either English sabins or metric (SI) sabins.

For some sound-absorbing elements, such as people, it is most convenient to quantify their sound-absorbing capabilities using sabins directly; the typical human being, fully clothed, provides about 5 sabins or sound absorption units at midfrequencies (500 Hz and 1000 Hz). Large people might have large surface area and be worth a few more sabins and children might provide a bit less, but 5 sabins per person is a good average.

In a theater or church, we look at the absorption of the audience in a number of ways. One way is just to count the number of people and assign about 5 sabins per person; this will tell you approximately the total sound absorption units in the audience. Another way is to use standard published data for absorption coefficients of a seated audience. For example, in the midfrequencies, typical absorption coefficients for occupied seating are about 0.85. Note that this is similar to the value of 5 sabins per person squashed out over a typical audience seating area of 6 ft² or 0.56 m² (seats approximately 2 ft or 0.6 m wide, rows 3 ft or 0.9 m on center). The absorptive properties of musicians on a stage may be different from those of an audience. The performers absorb more sound per person, not because of their musical talents, but because they are spaced apart (with around 15 to 20 ft², or 1.5 to 2 m² per musician) and offer more surfaces than the densely packed audience. Audience seating shows an interesting phenomenon of its own at low frequencies, caused by the grazing of sound over the periodic spacing of rows. The seat-dip phenomenon, whereby additional absorption of 10 to 20 dB is introduced at low frequencies around 150 Hz, is sometimes evidenced even from 100 to 400 Hz. Even the air in a theater absorbs sound. The amount of air absorption changes with humidity; drier air, typical in winter or in air-conditioned areas, absorbs more sound energy than more humid air.

Absorption coefficient data (such as those listed in Table 2.1 of this book) can clearly show the trends and patterns of absorption performance for materials. For example, coarse block can be quite porous on the surface and absorb about 30 percent of sound at all frequencies; however, when the block is painted, the pores are sealed and absorption plummets. This same degradation of performance occurs when acoustic tile is painted. The absorption values for fabric cover a wide range, but the trend is that heavier fabrics (which may offer greater flow resistance), when bunched up to increase the available surface area and spaced away from the wall, show improved values at all frequencies compared to lighter fabrics stretched tight close to the wall. Glass, gypsum board, and plywood paneling all show the characteristic panel absorption peak at 125 Hz or so. Large openings, such as the stage, balcony overhangs, and ventilation grilles, all have open areas that absorb sound. Unoccupied unupholstered seats absorb far less sound than occupied or upholstered seats, and empty upholstered seats have nearly the same absorption characteristics as when they are occupied—hence the advantage of upholstered seats in theater design to stabilize the reverberation time of a room, empty or full.

Quieting the Noisy Restaurant

BY ROBERT B. NEWMAN

Several years ago, an elegant restaurant in a large sports building was fully renovated. The new dining room was beautifully finished with a carpeted floor, good furniture, and a handsome ceiling of natural oak boarding. But, as soon as the restaurant opened, it was painfully apparent that this dining room was an acoustical disaster. Even when partially occupied, it was intolerably noisy. What could be done?

An acoustics consultant was called in, and he recommended covering the 5000 ft² of oak ceiling with sound-absorbing material. "Cover our beautiful oak ceiling? But we thought that wood was good for sound." "If you want to have reasonably quiet conditions, you'll have to cover it with something that absorbs sound." So, at considerable expense, the owners installed a fabric-faced fiberglass board treatment over the oak boarding and then covered this with narrow oak slats, spaced 6 in apart. The extensive remedial work wasn't cheap, but it solved the problem dramatically, providing a dining room that is now pleasant and reasonably quiet.

What qualities make a restaurant pleasant? Many things, of course. But one thing that can certainly make it unpleasant is too much noise. On the other hand, if the space is too quiet, it can be unpleasant because diners can overhear other conversations and know that they are being overheard as well. There are conditions, however, between these extremes that most people would judge to be pleasant.

Restaurant Noise Sources

The level of noise considered satisfactory in a restaurant is much higher than people would find acceptable in an office. In a dining space, the noise one hears is made up of a combination of noise from people talking and eating, the ventilating system, the activity of waiters serving and cleaning up, and sounds from the kitchen. Sound-absorbing treatment is needed to quiet this noise. But, in fact, it is not only the absolute level of the noise in a restaurant that makes diners think the space is noisy; a restaurant is generally thought to be adequately quiet if people can easily converse with their dining companions without being bothered by activity noise and the conversation of others. Surprisingly, some "noisy" restaurants have been improved merely by using somewhat smaller dining tables. This reduces the distance between talkers at a table, making conversation easier. People can speak more quietly and, with smaller tables more widely separated, adjacent diners are also less disturbing. In order to achieve the same degree of quiet, you can imagine how much separation would be needed between adjoining tables in a dining hall filled with large circular tables for eight.

Restaurant noise is typically between 20 and 30 dB higher than office noise—four or five times as loud. In an office, one must use the telephone, carry on conferences, and work with freedom from distraction. People are usually farther apart than in a restaurant. They may also be separated by screens or partitions, and there isn't constant conversation at every desk.

In a restaurant, with the closer spacing of people and without dividing screens, the background sound is largely controlled by the level of conversation of many diners, resulting in much more noise than one gets in an office. If the space is finished with hard, sound-reflecting materials, the levels of speech noise are likely to be so high that people must almost shout to each other to be heard. This makes the room noisier and noisier, and the environment becomes very unpleasant, although there is certainly complete privacy between tables in such a noisy setting.

By contrast, if diners arrive early in a sparsely occupied, quiet dining room that has been amply treated with sound-absorbing materials, they find they must talk at a very low level to avoid being overheard by everyone else in the room. As the room fills, however, conversation becomes relaxed and the surroundings become more comfortable.

Importance of Ceiling Treatment

The most important surface in the room for the control of noise is the ceiling. If the ceiling is hard and sound reflecting (wood, plaster, or concrete), sounds spread everywhere with little reduction, and noise levels build up. If the ceiling is sound absorbing, much less sound is reflected to other areas, and the sound level falls off as one moves away from the talker.

The ceiling material in a dining room should be an efficient sound absorber, having a Noise Reduction Coefficient (NRC) of at least 0.65. The NRC rating is an average sound absorption value commonly used to describe the general efficiency of a sound-absorbing material. Equally important, however, is the absorption coefficient of the ceiling treatment at around 2000 Hz, the frequency where there is the greatest contribution to speech intelligibility. At 2000 Hz, the absorption coefficient should not be less than 0.80 to absorb as much as possible in the range where speech intelligibility is carried. Many of the commonly available acoustical materials will meet this specification. One is not limited to standard acoustical tiles. Installations can be made up with fiberglass panels faced with fabric, wood slats, perforated metal facings, or grilles—in great variety—and these treatments will achieve the desired performance.

The essential element of any sound-absorbing treatment is the sound-absorbing core. The sound-absorbing element should be at least ¾ in thick, preferably a glass fiber product—these are the most efficient. Wood strips or a rough-textured material alone will provide no sound absorption. Acoustical plaster is virtually worthless for providing sound absorption.

We sometimes find that a restaurant may be considered quite satisfactory when it is half full, but intolerably noisy when full. This situation is almost always the result of inadequate sound absorption in the space. Usually the ceiling is hard, and absorption must be added before it will be acceptably quiet for full occupancy.

Despite the importance of the need for good sound absorption on the ceiling, a variety of conditions can produce a pleasant and satisfactory dining environment. Every reader will probably think of a favorite restaurant that does not have a sound-absorbing ceiling. One of my favorites is in a large hotel. It has an ornamented plaster ceiling, carpeted floor, large white damask tablecloths, fabric-upholstered chairs, heavy draperies, and a pianist. The tables are widely spaced. Thus, there is a low density of diners and the general level of conversation noise is reasonable—but, when the pianist stops, you can hear other conversations. A sound-absorbing ceiling might make the music unnecessary, but the room is pleasant the way it is.

Carpet on the floor reduces the noise of chair scraping and dropping objects. Tablecloths reduce the clatter of dishes and cutlery and also add a bit of local sound absorption. Draperies and wall treatments are effective when they are near the diners, but they have little effect on the general noise level in a wide room. Acoustical treatment on the ceiling is the best treatment since it distributes absorption throughout the space and reduces sound reflections off the major surface in the room. Also, because of the ceiling's inaccessibility, many effective materials may be considered that would not be durable enough to use on the walls or floor.

Kitchen Noise

Although the noise of other diners talking is the principal problem in most restaurants, there are occasions when kitchen noise can be annoying, especially to people seated near the serving doors. There should always be doors at the entrance to the kitchen or, at least, a circuitous passageway between the kitchen and dining area, finished with sound-absorbing materials.

To reduce the kitchen activity noise at its source, a sound-absorbing ceiling in the work area is very helpful. While this need not be quite as efficient as the ceiling in the dining area itself, it should have an NRC of 0.55 or better. Attention should also be given to such machines as dishwashers and ventilation fans, which should be resiliently mounted so that they do not vibrate the structure and radiate noise.

On several occasions, after using every conceivable means of quieting a dining area in the design of a new restaurant, an owner has expressed dismay at the level of noise heard during a busy evening. When receiving this sort of criticism, designers should remember that even outdoors at a picnic, the noise level of a hundred people can be surprisingly loud, even though they are surrounded by the ultimate in sound absorption, the open countryside.

Vibration Control Design
of High-Technology Facilities

BY ERIC E. UNGAR, DOUGLAS H. STURZ, AND C. HAL AMICK

High-technology equipment, such as that used for the production of advanced integrated circuits, for precision metrology, and for microbiological or optical research, requires environments with extremely limited vibrations. Ground motions, personnel activities, and the extensive support machinery typically present in high-technology facilities may produce unacceptably severe vibrations, unless mitigation of these vibrations is taken into account in the facility design. This discussion is intended to present pertinent facility design criteria and to summarize approaches to achieving the desired vibration environments.

The first of the following sections discusses the development of simple, practical facility criteria from equipment specifications and presents a suggested set of general criteria. Subsequent sections present an overview of approaches and means for dealing with vibrations generated by sources outside and within a facility.

Vibration Criteria

A completely vibration-free environment is as unachievable as are such other idealized abstractions as immovable objects, irresistible forces, or perfect vacuums. Fortunately, in practice it generally suffices to provide an environment that is adequately vibration free—that is, an environment that does not exceed suitably selected vibration limits. Establishment of appropriate limits is crucial to the successful design of a sensitive facility; limits that are insufficiently stringent lead to degradation in the performance of sensitive equipment, whereas limits that are too stringent lead to excessive complexity and increased costs.

For a given sensitive facility—be it a plant area for the manufacture of integrated circuits, a microbiology research laboratory, an optical calibration facility, or a metrology laboratory—it is logical to select the limit of permissible vibration to correspond to the most severe vibration environment under which all critical items of vibration-sensitive equipment can operate satisfactorily. This approach to development of an appropriate vibration criterion is relatively straightforward, at least in concept, if all equipment items to be placed in the facility are fully identified and if the acceptable vibration limits for all items are known. In this case, one simply needs to require that the environmental vibration in each given frequency band does not exceed the greatest vibration magnitude that is acceptable for the equipment with the most stringent limitation in that band. (Of course, different equipment items may determine the criterion values in different bands. Also, different criteria may apply in different locations of a given facility.)

In practice, however, development of facility vibration criteria involves some complications. The equipment to be installed in a facility may not be fully defined at the time the facility is being designed, and new equipment with initially undetermined sensitivities is likely in the future to replace or supplement the originally installed equipment. Furthermore, acceptable vibration limits for many sensitive items of equipment are not known adequately. Although most equipment manufacturers provide some sort of specification that sets limits on the environmental vibrations of the areas where this equipment is to be installed, many of these specifications tend to be overly conservative and many are inadequately defined in that they do not indicate the frequency ranges in which they apply or the frequency bandwidths in which measurements are to be made.

FACILITY CRITERIA

Figure T.8 lists velocity criterion values (corresponding to the 8- to 80-Hz region) that have been found suitable for facilities housing various classes of sensitive equipment for sev-

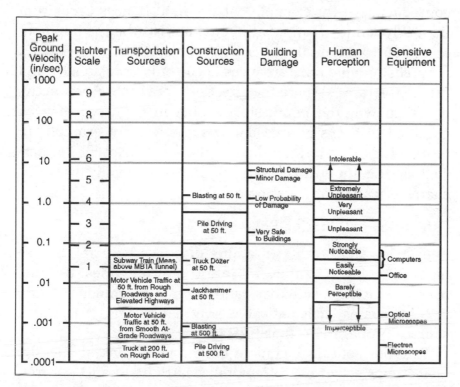

Figure T.8 General references for vibration levels.

eral different building space usages. It is important to note that the listed criterion values were developed on the basis of available equipment data; although these criteria are generally conservative and have led to numerous successful facility designs, it is possible that they may not be adequate for some particularly sensitive new items of equipment.

In selecting a vibration criterion for a facility, one needs to consider the extent to which occasional disturbances may be acceptable. For example, the blurring of an image in an optical microscope resulting from occasional heavy foot traffic in a nearby corridor may annoy the microscope's user a little, but is likely to have little effect on the progress of research in the laboratory, whereas even a brief disturbance of a manufacturing process may lead to serious production losses. Clearly, one would be inclined to prescribe stringent criteria in situations where disturbances have more severe consequences—for example, where disturbances occur continuously and where continuous undisturbed equipment operation is essential—whereas one might relax the criteria for areas in which disturbances occur only occasionally and can be tolerated. Where the effect of vibration on productivity is known, the permissible vibration exposure may be specified in terms of exceedance statistics—for example, in terms of the L_n vibration levels (the levels exceeded n percent of the time) analogous to those commonly used for characterization of environmental noise.

Categories of Vibration Sources

The major sources of vibrations of concern in relation to high-technology facilities fall into three categories: external sources, internal activities, and service machinery. Figure T.9 illustrates schematically how vibrations from such sources propagate to sensitive areas.

External sources include ambient vibrations at the site (sometimes called microtremors), nearby road and rail traffic (including underground and elevated roads and rail sys-

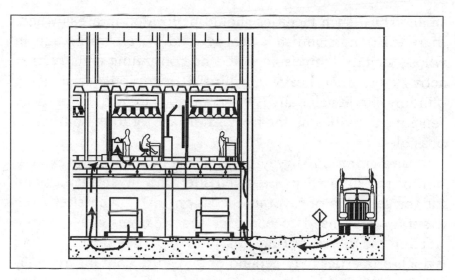

Figure T.9 Propagation paths for ground- and structure-borne vibrations through buildings.

tems), construction activities (including blasting), and machinery operating in the vicinity (either outdoors or in nearby buildings).

Internal activities include personnel walking (footfalls) and service activities (e.g., repair and construction), in-plant vehicles (such as forklifts and carts), and production work (e.g., actuation of production machines or other tools).

Service machinery includes all mechanical and electrical equipment that either is part of the building's system or that is installed by the building's users. It includes air-conditioning and distribution fans, chillers, cooling towers, furnaces, and all pumps, compressors, and vacuum pumps, as well as elevators and mechanically actuated doors and loading platforms.

Externally Generated Vibrations

SITE SELECTION

It stands to reason that vibration-sensitive facilities should be sited where ambient ground vibrations are acceptably small. Thus, such facilities should be located in areas where there exists no significant nearby road or rail traffic and in whose vicinity there is expected no continuing construction activity or other heavy machinery operation. At a given building site, it generally is advantageous to locate vibration-sensitive activities as far from external vibration sources as possible.

A site vibration survey generally is advisable for evaluation of the suitability of a given site for a given facility and for the selection of favorable locations at a given site. Such a survey may need to consider the weather-related variations in the vibration-transmission properties of the ground (which properties may depend on moisture content, temperature, and the height of the water table, among other things), local geological nonuniformities (e.g., variations in depth of bedrock), the types of foundations to be used, and the daily variations in traffic.

REDUCTION OF VIBRATIONS FROM TRAFFIC

In some instances, one may be able by design to reduce the vibrations generated by an external source. For example, it is well known that the most severe vibrations associated with road traffic result from heavy vehicles moving rapidly along roads with surface irregularities. Thus, one may reduce vibrations at a site by keeping heavy trucks away from sensitive facilities, limiting the permissible speeds, and smoothing the road surface. Certainly, speed bumps, potholes, misaligned slabs, and expansion joints (in bridges) should not be permitted near vibration-sensitive facilities. Similarly, if the related costs are acceptable, one may consider replacing jointed rail by continuously welded rail in railroad tracks passing near sensitive facilities, and/or placing such rail on thick ballast beds or on resilient rail support systems.

In some situations, schedule control may be most cost effective. For example, one might confine the use of critical electron microscopes to nighttime, when nearby construction activities have ceased and road traffic is minimal. Alternatively, one might consider installing a vibration-monitoring system that senses approaching trains and halts the operation of sensitive equipment during train passages.

PROPAGATING VIBRATIONS

We are aware of no practical means for shielding facilities from vibrations that propagate along the ground. Berms, heavy walls, and other structures above the ground have very little effect on ground vibrational waves at the frequencies of primary concern here. The same is true of trenches, sheet piling, slurry walls, and similar underground structures or geotechnical means (e.g., grout injection) of practical size, largely because the wavelengths at the frequencies of concern tend to be great (of the order of 100 ft) and discontinuities that extend over only a fraction of a wavelength fundamentally can provide little attenuation.

FOUNDATION DESIGN AND ISOLATION

Some vibration control benefits can be obtained from appropriate foundation design and from isolating key parts of a facility from soil vibrations. In situations where the ambient vibrations of the bedrock are of relatively small magnitude compared to those of the surface soil, one may base the building foundations on the bedrock and avoid their coupling to the surface soil (see Figure T.10). On the other hand, where the surface soil vibrates less than bedrock, mat foundations or spread footings are preferable from the vibration standpoint. Which of these situations exists depends on the soil conditions and on the locations and types of the predominant external vibration sources.

In some instances, one may also be able to design "tuned" footings that reduce the vibration intrusion in certain frequency ranges. The footing and adjacent soil together act like a spring that supports a portion of the building's mass, so

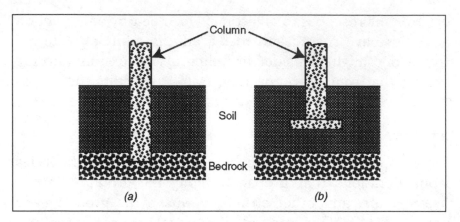

Figure T.10   Support of a building on quieter stratum: *(a)* column to bedrock; *(b)* spread footing on soil.

that this arrangement acts somewhat like a classical spring-mass system. At frequencies that lie considerably above the system's natural frequency, only a small fraction of the ground vibration amplitude is transmitted to the mass—that is, to the building structure. One may tune the footings (generally by adjusting their footprint areas and shapes) so that the desired attenuation is obtained in the frequency ranges of interest—or at least to avoid resonances at frequencies at which relatively severe ground vibrations are present. In designing such tuned footings, one generally needs to consider the frequency distribution of the ground vibrations in three orthogonal directions, and one needs to account for the different footing stiffnesses and building components that relate to motions in the different directions.

Where footing design cannot provide sufficient attenuation, selected parts of the building may be isolated from ground vibrations by supporting their columns or footings on resilient elements, such as neoprene bridge-bearing pads or "air mounts" (pneumatic springs), as shown in Figure T.11. The selection of these elements depends on the frequency ranges that are of primary concern. Here again, these resilient elements and the mass they support act like a

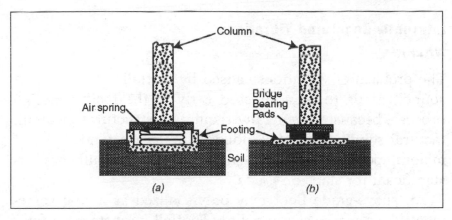

Figure T.11 Examples of resiliently supported building columns.

mass-spring system, which attenuates vibrations above its resonance frequency, but amplifies vibrations near that frequency. Such a specially designed system has the advantage that it can be made very resilient, with a resonant frequency below the range of concern, and with the potential for providing considerable attenuation of intruding ground-borne vibrations. It has the disadvantages of considerable construction complexity and attendant costs as well as the potential for increasing vibrations resulting from internal activities.

Figure T.11 illustrates schematically how building columns may be supported on air springs or bridge-bearing pads, but does not show the lateral restraint that must be provided at the column bases to ensure stability of the structure, particularly with regard to earthquakes. Placement of air springs in a pocket that is open on one side permits removal and replacement of the air springs in the event that becomes necessary. One would merely need to deflate the air spring, permitting the column footing to rest on the lips of the support; one could then slide the air spring out, replace it, and inflate the replaced air spring until the column footing is again supported only on the spring.

Internally Generated Vibrations

WALKING

The problem of vibrations caused by footfalls (walking personnel) needs to be addressed early in the facility design process, because it generally requires a structural or architectural solution. Footfall-induced vibrations tend to be of major importance for above-grade floors, but usually are less significant for slabs on-grade.

An above-grade floor may be visualized as acting somewhat like a massive trampoline. Footfall impacts on a floor set the floor structure into motion, subjecting any equipment resting on it to corresponding vibrations. Footfalls near the center of a bay tend to cause the greatest vibrations, and the vibrations always tend to be most severe at midbay and least severe near columns. Thus, footfall-induced vibrations and their effects may be reduced by confining heavily traveled areas (e.g., corridors) to regions near column lines, placing sensitive equipment near columns, and keeping as much distance as possible between heavily traveled areas and sensitive equipment.

It is also well known that rapid walking causes more severe footfall impacts than slower walking and that several people walking in step can cause very severe vibrations. The probability of obtaining such conditions may be reduced by avoiding layouts with long, straight corridors that permit rapid walking and by instituting administrative measures (e.g., posting signs and educating personnel to walk slowly).

However, the most reliable solutions to the footfall-induced vibration problem usually are structural, consisting of (1) making the floor structures stiff enough so that the footfall-induced vibrations associated with expected foot traffic remain within acceptable limits and/or (2) separating the structures on which people walk from those that support the sensitive equipment.

Effective control of footfall-induced vibrations by structural design in essence consists of providing a sufficiently stiff structure. Floor structures with a high degree of stiffness can generally be obtained by using small column spacings; otherwise, considerably deeper than usual floor girders, joists, and slabs may need to be employed.

It often is convenient to separate areas exposed to foot traffic from adjacent areas that house sensitive equipment by providing a separation between these areas and, ideally, designing the structure so that these areas do not share a common support (girder or column line). The separation should consist of a physical joint, which may include resilient supports and seals. In some situations it may be useful instead to provide "bridges" on which people can walk, where these bridges (which may be at the level of a raised floor) are constructed so that they are supported only at the columns, without making contact with floors on which sensitive equipment is supported.

IN-PLANT VEHICLES

The aforementioned concepts for controlling footfall-induced vibrations also are useful for limiting vibrations due to in-plant vehicles. In addition, because a vehicle entering on a floor slab or leaving a slab in effect produces a suddenly applied load, it is desirable to reduce the suddenness of load application—for example, by using joints with long, interlacing fingers or by having joints arranged so that only one wheel of a vehicle at a time crosses the joint. It is also useful to use soft pneumatic tires on all vehicles and to keep the surfaces traversed by the vehicles smooth and free of surface discontinuities.

PRODUCTION MACHINES AND ACTIVITIES

The effects of vibrations resulting from production-related machines and activities that may cause disturbances (such as machine adjustments or the installation of gas cylinders) can be reduced by keeping these as far from sensitive equipment areas as possible, by locating these machines and activities in areas where the supporting structures are relatively stiff (e.g., near columns), and by supporting machines on resilient vibration-isolating systems. In general, the same vibration control concepts that are discussed in the following paragraphs in relation to service machinery apply here also.

Machinery Vibration

MACHINERY SELECTION

It is useful, where possible, to select alternative mechanical and electrical equipment types that inherently are relatively free of vibration. For example, rotating compressors tend to produce considerably less severe vibrations than reciprocating compressors, because their inertia forces are better balanced; for the same reason, multicylinder (particularly opposed-piston) engines and compressors are preferable to single-cylinder machines. Similarly, it is advisable to choose the better balanced of two otherwise similar machine models, and one may do well to opt for the purchase of equipment with the best economically feasible field-balance specifications; however, we have found that ultrafine balance usually is unnecessary.

MACHINERY PLACEMENT

It is advisable generally to keep as much distance between vibration-sensitive equipment and vibration-producing machinery as possible, to support vibration-producing machinery on stiff structural components, and to provide this machinery with efficient vibration isolation systems. Machinery isolation usually involves the use of well-known approaches and readily available technology; however, special care generally is required to avoid bridging of the isolation via piping and conduits. Good planning and facility layout usually go a long way toward minimizing the rerouting of pipes and ducts and their special isolation that may otherwise be required for vibration control.

MACHINERY ISOLATION

Figure T.12 illustrates three approaches to reducing the transmission of vibrations generated by heavy equipment located at grade to the adjacent soil (and thus to the building). Where the building rests on footings that do not communicate directly with bedrock, it is useful to let the machinery act on the bedrock via suitable columns or piers that do not make direct contact with the soil. The other approaches involve relatively straightforward isolation of a machinery base from the building floor slab or ground. Two alternatives to the conventional use of steel coil springs are illustrated in Figure T.12*b* and *c,* an arrangement in which a layer of fill serves as the resilient isolation element and a system that employs air springs in a pocket arrangement (for easy replacement) to provide very effective isolation.

It should be noted that a simple cut in a concrete slab that rests on soil (or a gap between two structural components, both of which rest on soil) provides no significant reduction in the transmitted vibrations. The dynamic properties of concrete and of soil are sufficiently alike that the vibration in the frequency ranges of general concern is transmitted around the gap via the soil with very little attenuation.

In high-technology facilities, as in any complex dynamic system, careful attention needs to be paid to a large number of details in order to ensure that the desired vibration performance is indeed obtained. All potential vibration transmission paths that may short-circuit machinery isolation systems or structural breaks need to be

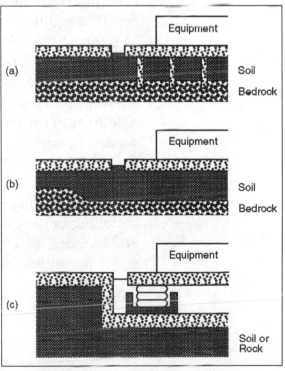

Figure T.12 General designs for isolating vibrating equipment from buildings.

considered and eventually treated. This includes piping, ducts, and conduits that may bridge the isolation systems and gaps, as well as such auxiliary structures as partitions and pipe racks.

Isolation of Sensitive Equipment

It is generally advisable to locate vibration-sensitive items in areas where vibrations due to external sources and internal activities are comparatively benign. Such areas typically include on-grade locations that are away from external traffic and not too close to mechanical equipment (e.g., elevator machinery or pumps, furnaces, and air handlers that may be in mechanical equipment rooms). Favorable locations on upper floors typically are areas near columns and major girders, as far as possible from heavily traveled corridors. Confining placement of sensitive equipment to such favorable locations or to limited, specially designed areas can result in significant structural savings.

Supporting vibration-sensitive equipment on raised "computer" floors is generally undesirable, because vibrations produced by people walking on such floors can be transmitted relatively readily to the equipment. It is preferable to support the equipment from the structural floor via separate pedestals or via separate, braced sections of raised flooring that do not make direct contact with the portions of the flooring on which people can walk. Bracing of raised floors designed specifically for vibration control (sometimes including bolting of floor tiles) has been found to be considerably more effective for the protection of equipment from small amplitude disturbances than has the bracing commonly used for code compliance in seismically active areas.

Many items of sensitive equipment include extremely resilient internal isolation of their critical components, and some can be ordered with special isolation tables or cradles. For this reason, supporting such an item on an additional isolation system typically provides little benefit unless this isolation system incorporates extremely flexible elements. Two-stage isolation, involving a heavy base on soft springs under an equipment item incorporating resiliently supported elements, may be useful on occasion; however, very soft springs and large masses are usually required to provide significant isolation in the frequency ranges of concern.

In some instances, one may need to protect sensitive equipment located at grade from ground vibrations. In such cases, in view of the predominance of disturbances at relatively high frequencies, conventional isolation arrangements or those shown in Figure T.12*b* and *c* merit consideration.

Cost-efficient design of a high-technology facility from a vibration standpoint generally requires collaboration of a vibration control specialist with the facility's user and architect (to develop favorable layouts), with geotechnical and structural engineers (to arrive at desirable footings and structural configurations), with mechanical engineers (to obtain adequate vibration isolation and noise control for air-handling and other service equipment and the related piping and ducts), and often with process and other equipment specialists. Involvement of a vibration control engineer beginning with the early design stages usually is most beneficial, because in the early stages there often exist opportunities to make design choices that are beneficial from a vibration standpoint, but which imply little or no increased cost.

Good design from the start is important, but even the best design is useless unless it is implemented properly. For this reason, it is advisable during the entire design and construction process to monitor the myriad details that may affect vibration by careful review of relevant design, construction, and shop drawings, and by repeated field inspection in the course of construction. Ideally, vibration measurements should also be performed after the facility is completed so that conformance with vibration specifications can be verified and any residual problems can be identified and resolved.

Sound Masking: The Results of a Survey of Facility Managers

As part of a market research project, a questionnaire on sound masking was sent to approximately 1000 facilities managers throughout the United States.* The 405 questionnaires that were returned represent about 140 million ft² (13 million m²) of office space. The effort yielded some interesting insights into the prevalence of sound masking, common attitudes about the technique as a way to improve speech privacy, and the way in which the decisions are made regarding its installation.

The Use of Sound Masking

A broad spectrum of U.S. industry is represented in the study sample. Dividing the returns into nine general industrial categories, the use of sound masking can be summarized as follows:

*This research was carried out by the Concord Consulting Group, Concord, Massachusetts.

231

Industry	Number of Sites	With Masking
Computers, data processing	71	25%
Finance, insurance, real estate	60	25%
Entertainment, education	23	30%
Services	42	26%
Infrastructure	53	42%
High tech	29	31%
Retail/wholesale	13	23%
Manufacturing	58	31%
Other	56	36%
Total	405	32%

In general, the industries show about the same enthusiasm for sound masking (a small spread in the percentage of sites reporting sound-masking installations). The exceptions are the "infrastructure" (including energy, transportation, utilities, and telecommunications) and "other" industries—both with above average adoption rates of sound masking.

Offices of different sizes were found to have different sound-masking adoption patterns:

Size of Site (Ft² of Open Plan)	Extent of Sound Masking	
	% of Sites	% of Area
0–10,000	15.4	8.7
10,001–50,000	17.1	12.4
50,001–100,000	23.6	9.8
100,001–500,000	34.7	17.2
500,001 and above	44.4	7.2

A change in the rate of installation of sound-masking systems would be a good indication of changes in the acceptance of the technique. Sixty-eight percent of the sound-masking users provided information on when their sound-masking systems were installed. The survey sample indicates a clear upward trend in the use of sound masking.

Market Environment

The survey showed a generally favorable market environment for sound masking. Counting all of the respondents:

84 percent are familiar with sound masking, 49 percent have evaluated sound masking, 32 percent have installed sound masking, and 24 percent plan future sound-masking systems.

It is significant that almost half of the sites reporting (all of those with sound masking, and some of those without) have evaluated the technique as a potential solution to office speech privacy problems, and nearly a third of all the sites reporting have installed at least some sound masking. The smaller fraction of respondents that reported future plans for sound masking does not indicate a downturn in the market—many respondents answered with regard to specific projects and not with regard to general intentions over the next 10 years or so. Further analysis of the data showed that 51 percent of those who already have sound masking plan to expand their use of it, and 16 percent of the sites currently without sound masking have plans to install it for the first time.

A qualitative sense of the distribution of new or additional sound-masking installations (sites, not total area covered) can be gained from the survey. Facility managers were asked in what types of offices sound masking would be installed, and if it would be installed in "all," "most," "some," or "none" of the designated space.

Fully 42 percent of the facility managers planning on future installations of sound masking in open-plan offices indicated that it would be installed in all of their new office space of this type. The survey also suggests that sound masking will be more often applied to full-wall offices and public spaces than it has been in the past.

The Open-Plan Office

The open-plan office is key to the sound-masking market. Rumors about the demise of this type of office were put to rest by this survey. Nearly all of the respondents shared their future plans with regard to the open-plan system:

- Five percent do not plan to install any open-plan offices in the future.

- Six percent plan to use open-plan offices less than in the past.

- Seventy-five percent plan to use open-plan as much as they have in the past.

- Fourteen percent expect to utilize open-plan offices more than in the past.

Measured by the number of responses, the overwhelming majority of sites will continue their current practices of using open-plan offices. The few reports of plans to reduce or eliminate the open plan are more than offset by reports of intentions to increase the use of the open plan.

Why Is Sound Masking Installed?

The majority of respondents (67 percent) indicated that sound masking was installed during refurbishing of office space, or the construction of new office space, in anticipation of an acoustic problem. An additional 15 percent of the sound-masking installations were initiated to fix a problem discovered after these new facilities were occupied. Therefore, about 82 percent of all sound-masking installations are done in conjunction with construction or renovation. Only 13 percent of sound-masking installations are made in order to "fix" a speech privacy problem in existing facilities.

The most common other reason for installing sound masking was to provide security (3 percent).

Sound-Masking Effectiveness

Among the facility managers that have sound masking installed, the strong, but not universal, consensus was that sound masking is an effective technique. There was a wide variation in the perception of exactly how effective the systems were.

The data did not allow a rigorous correlation between the perception of effectiveness and other aspects of the sound-masking systems. However, in the context of the other answers and comments on the questionnaires, it is speculated that this variation is due in part to the degree of sophistication brought to the sound-masking system design and installation.

Designers/Installers

Once the decision is made to go forward with a sound-masking system, there are a variety of sources for design and installation services. The facility managers indicated that in the future the allocation of this work will follow roughly the patterns of the past. Specifically, the respondents indicated that for future systems, the breakdown is as follows:

Source	Design	Install
Architect/designer	13%	0%
Acoustic consultant	40%	10%
Sound system vendor	24%	45%
Security system vendor	0%	2%
Electrical contractor	0%	17%
In-house staff	15%	12%
Other	0%	2%
Don't know	8%	12%

The estimate that about 40 percent of the facilities depend on an acoustic consultant for the design of sound-masking systems probably understates their importance. Some acoustic experts operate as invisible consultants working for the architect or the designer—the latter being given credit for 13 percent of the design effort.

As would be expected form the economics of the situation, sound system vendors dominate the installation process.

Unexpectedly, the in-house staff represents a significant factor in design and installation, overshadowing the security system vendors and challenging the electrical contractors.

Barriers to Adoption

About 25 percent of all respondents said that they had evaluated sound masking for specific areas within their facilities and had decided not to install the systems. This can be broken down further to show that 26 percent of the facilities that do have sound masking somewhere at the site have rejected the system for other specific areas. And 25 percent of the facilities that still do not have any sound masking have at least considered the technique before finally rejecting it.

A number of respondents evaluated the barriers that led to the rejection of specific sound-masking installations. The three major barriers cited were (1) the cost of the system, (2) the conclusion that the speech privacy problem was not too severe after all, and (3) the feeling that sound masking would not be an appropriate solution. These three issues are obviously intertwined in the complicated decision process. Unfamiliarity with the technique and architectural limits were not cited as equally serious barriers.

Alternatives

Where sound masking was considered, but not installed, other courses of action were reported. The alternate methods of solving speech privacy problems were as follows:

- Fifteen percent changed the design of the partitions.
- Twenty-one percent installed full walls.
- Fifteen percent changed ceiling tiles.
- Twenty-one percent used a variety of other techniques.
- Seven percent plan to hire a consultant.
- Twenty-one percent took no action.

The large number of "no action" responses supports other indications that many organizations decide to ignore speech privacy problems and accept the loss in productivity. On the other hand, where action was taken to improve the situation, the alternatives were, in general, expensive: install full walls and change the partitions. Among the most-cited other options were attempts to change the behavior of the office occupants!

Summary of Comments

The questionnaire invited unstructured comments—96 people responded. The statements ranged from cryptic fragments to well-thought-out observations on sound masking, at all levels of sophistication.

THE CHANGING FUTURE OF SOUND MASKING

Respondents suggested that, in the future, speech privacy will be of growing concern. There will be more potential for distraction because of the use of voice mail, intercom telephones, and so on. In addition, "better" furniture and panels will make the background of the open-plan office quieter.

One of the most interesting suggestions was that there would be a change in the character of the speech privacy problem as voice recognition systems come into common use.

SOUND MASKING IS NOT ALWAYS THE TOTAL SOLUTION

It was widely observed that sound masking is a powerful technique, but the best solutions to speech privacy problems required the simultaneous optimization of a variety of elements—carpets, panel materials, panel configuration, ceiling and wall treatments, even visual privacy.

In some instances the speech privacy problem had to be solved, and sound masking was not selected. Many of the responses indicated that this is a major economic decision. The most often mentioned stand-alone alternatives are even more expensive—change to full walls, create more closed conference rooms, change the open-office panel system, and so on.

Effectiveness of Sound Masking

As expected, many respondents volunteered enthusiastic endorsements of sound masking—some to the effect that no open-plan office should be without it.

On the other hand, a number of respondents expressed some dissatisfaction with their sound-masking systems. A few even reported the discontinuation of sound masking because of negative experiences. This dissatisfaction seems to stem from failure to achieve the expected performance either because the designer did not accurately characterize the expected effects or because of a lack of sophistication in design, installation, and/or use. Some of the negative commentary was simply grumbling about open-office systems in general.

The Sound of NC-15

BY ROBERT S. JONES

In general the acoustical environment goal for concert halls is NC-15. But what does NC-15 mean and what does it sound like? To answer the first part of the question please refer to Figure T.13, which shows the universally accepted family of noise criteria (NC) curves. These are used to acoustically rate various types of space in many kinds of building. They have been established by plotting decibel (dB) levels at each octave band center frequency from 31.5 to 8000 Hz for each NC curve.

As you will notice, the NC-15 curve is the lowest, with the exception of the heavy curve labeled "Approximate Threshold of Hearing for Continuous Noise." The NC-15 curve, being the lowest, is therefore the quietest and most difficult to attain.

In answer to the second part of the question—what does NC-15 sound like?—consider this. Imagine that you are standing all alone at center-stage front, in a concert hall. All mechanical and electrical systems, as well as other utility devices, are operating. You are listening intently, yet practically no sound comes to your ears. NC-15 is like that, and

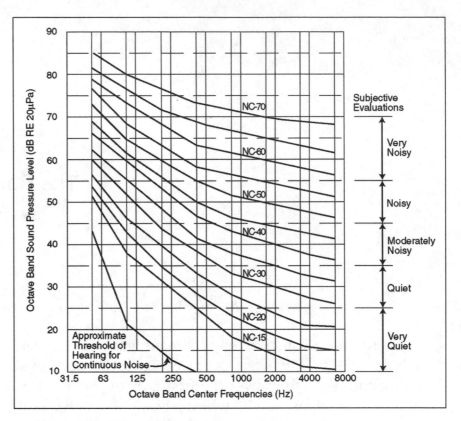

Figure T.13   NC curves used for rating interior noise environments.

that is quiet! You would hear no rumble from fans, no whoosh from the huge quantity of air being dumped into and returned from the hall. There would be no hum from lighting ballasts, no bulb filament buzz, and no sound during the dimming of lights. NC-15 is indeed quiet and you would be impressed!

Now we come to one very special part of concert hall design. How is it possible, with a complete chiller plant with pumps and with all those fans pushing 87,600 ft^3/min of air into and out of the hall and with all the other miscellaneous pieces of noisy equipment operating, to get NC-15 in the main hall? Such quietness does not come inherently in the design, nor does it come easily during construction of the building.

It really takes a team-coordinated effort to design, assemble, and properly install so many elements and components and not have many things go wrong that could spell near disaster for the acoustical goal of NC-15. So it was with the professional help and assistance of associated architects, mechanical consultants, electrical consultants, general contractors, the clerk of works, and many workers, together with the acoustical consultants, that make it possible to achieve what some have considered to be almost impossible.

For this specific hall, there are four main fan systems, each with its own supply and return fan, delivering air to and from the hall and an additional six systems serving the ancillary spaces around the hall, so that together there are 20 fans. These four main systems are located in a ring mechanical equipment room that very nearly surrounds the entire top of the concert hall. The fans supply air through sheet metal ducts to open air slots located in the ceiling and under the balcony fronts. The final connection from the sheet metal ducts to the air slots is made with sections of round flexible duct and a transition piece which changes from a round duct to the long narrow shape of the slot.

Air is returned from the back of the stage, the front of the stage, and around the rear of the hall at the different balcony levels through sheet metal ductwork. The returns at the rear of the hall are tall, narrow grilles about 3 or 4 ft from the nearest seats.

The ring mechanical equipment room is housed in a steel and concrete structure that is totally independent from the concert hall structure. There is, therefore, a complete structural separation between the two. Of course, this break must be bridged by piping, conduit, and ducts, but each such bridge is fitted with flexible, resilient connections. The flexible connections serve to prevent or minimize vibration and noise from being transmitted across the structural break.

In most cases there is a double masonry wall which separates this ring mechanical room from the hall. The inner mechanical room wall is set on the concrete floor of the mechanical room floor and the outer wall is the poured concrete wall of the auditorium.

Due to the fact that the ring mechanical equipment is so close to the hall and that the ducts leaving the room generally had to immediately turn 90° up, down, left, or right, it was decided to use elbow duct silencers. These are specially designed devices with glass fiber material inside for absorbing some of the fan noise. Previously, not a single concert hall in the world, as far as is known, has employed elbow duct silencers for controlling fan noise, although they have been available for some years. Considerable concern was expressed with regard to the use of these devices, so the manufacturer offered to set up a test facility at its plant to allow the acoustical consultants to witness and actually take part in a test to verify acoustical performance of the elbow silencers. They were tested, with and without air flow, and successfully performed to their acoustical rated values. The decision was made to accept and use them, and it proved to be a good decision.

In most cases, the silencers are positioned such that one leg of the elbow actually sits in the wall of the mechanical equipment room. Where this could not be accomplished, duct extensions were used and the elbow silencers were pulled away from the wall. The walls of these exposed sections of duct were custom-fabricated as part of each elbow by the silencer manufacturer, to provide a high degree of sound transmission loss (TL) capability, so that unwanted noise could neither leak in nor leak out of these extended sections. Noise could therefore not escape from the duct before it reached the silencer, nor, depending on the silencer location in relation to the fan, could noise enter the duct after the silencer and before the duct's penetration of the mechanical equipment room wall.

The return air is collected vertically from the different levels of the hall and breaks out horizontally into the ceiling space of the main lobby located directly below the ring mechanical equipment room. The return duct turns up and through the equipment room floor and directly into the return air plenum of the fan.

Each of these return air connections is fitted with a high-TL box, which is attached and sealed to the underside of the equipment room floor slab. Each return air duct entering one of these boxes is fitted with a standard (not elbow type), duct silencer selected for the degree of sound attenuation required to prevent noise breakout to the lobby from each respective duct run.

The entire package of high-TL return air boxes and silencers was designed and built by the same silencer–noise control manufacturer, so that the performance for the package became the sole responsibility of one agency. These elements performed to the predicted acoustical levels published in the manufacturer's certified data tables.

To maintain total separation of the mechanical equipment room floor from the exterior building glazing and support system, an open gap had to be provided between the exterior glazing of the building and the turned-up floor along the outside edge of the ring mechanical room. To prevent unwanted noise transmission from the equipment room through this gap and down to the lobby beneath, a means had to be provided for resiliently sealing the gap. Sheets of "loaded" vinyl were cut, fitted, and sealed along this gap and even around protruding structural members. This loaded vinyl is a glass fiber–reinforced material loaded with high-density powders for maximum noise attenuation; it is strong, durable, and flame resistant. The material used in this particular application has a surface weight of approximately 1.5 lb/ft^2.

One other rather special feature of the mechanical system design is the inclusion by the mechanical consultants of special custom-made acoustical enclosures for all the fans in the ring mechanical equipment room. These enclosures are referred to as "metal sandwich panels" and are fabricated with a solid sheet metal exterior, 4-in-thick glass fiber between the framing members, and a perforated metal inner skin. They are designed to provide maximum noise transmission capability, as well as a high level of sound absorption due to the perforated metal backed with glass fiber, all of which faces the noise source or sources. The enclosures were also fabricated by the same noise control manufacturer.

The sound levels in the mechanical equipment room are so quiet that one can easily be understood during conversation in a normal tone of voice only 3 ft from one of the enclosures. They also serve to maintain the lowest possible sound levels while all the units are running simultaneously and thereby reduce considerably the chance of noise intrusion into the hall or other sound-sensitive spaces.

A sound lock vestibule with two separate doors is installed at the two ends to the ring mechanical room. These vestibules provide a means for maintenance personnel to enter or leave the mechanical room while still keeping one door closed. This seal prevents unwanted noise from escaping down through the hall ceiling, which is open through various slots to the concert hall and balcony seating directly below.

It is really quite exciting to sit up in the highest balcony, when the hall is empty, and listen intently for a trace of sound from 20 fans that are rumbling away, some of which are just a few feet behind you, yet not being able to discern any audible mechanical sound.

During a performance, when the orchestra or individual performer and the audience are enraptured together for a breathless few moments of silence, not a sound is heard—a rare acoustical phenomenon. Described in technical terms, one has then heard NC-15.

Mathematical Expressions

The following essential mathematical definitions have been included for reference.

Relationship between frequency and wavelength:

$$c = f\lambda \quad \text{alternately,} \ \frac{c}{f} = \lambda \quad \text{or} \quad \frac{c}{\lambda} = f$$

where
c = speed of sound (in ft/sec or m/sec)
f = frequency (in Hz)
λ = wavelength (in ft or m)

Decibel definitions:

$$\text{Basic:} \quad \text{dB} = 10 \times \log\!\left(\frac{W}{W_{\text{ref}}}\right)$$

where
W = sound power
W_{ref} = reference sound power (1×10^{-12} watts)

$$\text{Sound pressure level:} \quad \text{SPL} = 20 \times \log\left(\frac{p}{p_{\text{ref}}}\right)$$

where p = acoustic pressure
p_{ref} = reference acoustic pressure (2×10^{-5} N/m^2)
at the threshold of hearing

Reverberation time:

$$\text{RT}_{60} = 0.05 \times \frac{V}{A}$$

where V = room volume in ft^3
A = total room absorption = $\alpha_1 S_1 + \alpha_2 S_2 + \ldots + \alpha_n S_n$

where
$\alpha_{1,2,\ldots,n}$ = absorption coefficients for different room materials
$S_{1,2,\ldots,n}$ = surface areas (in ft^2) of materials corresponding to α_n

$$\text{RT}_{60} = 0.16 \times \frac{V}{A}$$

where V = room volume in m^3
A = total room absorption = $\alpha_1 S_1 + \alpha_2 S_2 + \ldots + \alpha_n S_n$

where
$\alpha_{1,2,\ldots,n}$ = absorption coefficients for different room materials
$S_{1,2,\ldots,n}$ = surface areas (in m^2) of materials corresponding to α_n

Transmission loss:

$$\text{TL} = 10 \times \log\left(\frac{1}{\tau}\right),$$

where   TL = transmission loss
τ = transmission coefficient

References in Acoustics

The following books provide detailed information on the theory and mathematics involved in architectural acoustics.

Beranek, L. L. *Noise and Vibration Control,* revised edition. Washington, DC: Institute of Noise Control Engineering, 1988.

Beranek, L. L., and I. L. Ver. *Noise and Vibration Control Engineering, Principles and Applications.* New York: John Wiley & Sons, 1992.

Cavanaugh, W. J., and J. A. Wilkes. *Architectural Acoustics: Principles and Design* New York: John Wiley & Sons, 1999.

Egan, M. D. *Architectural Acoustics.* New York: McGraw-Hill, 1988.

Harris, C. M. *Handbook of Acoustical Measurements and Noise Control,* third edition. New York: McGraw-Hill, 1991.

For information regarding concert hall acoustics design, a complete reference is:

Beranek, L. L. *Concert and Opera Halls: How They Sound.* Melville, NY: Acoustical Society of America, 1996.

Glossary

The following provides definitions for common acoustical terms used in this book and in the field of acoustics.

absorption The conversion of sound energy into heat energy by penetration into porous materials or reaction with flexible panels.

absorption coefficient The ratio of energy absorbed to that incident upon a surface, usually denoted by the Greek letter α; α ranges between 0 and 1, with 0 indicating total reflection and 1 indicating total absorption (or no reflection).

acoustics The science of sound.

active noise control The electronic cancellation of sound energy by introducing a mirror image of a sound wave to the original sound wave in a confined area.

ambient sound level The sound level at a location arising from all sound sources in an area.

American National Standards Institute (ANSI) A voluntary federation of organizations in the United States, concerned with the development of standards.

American Society for Testing and Materials (ASTM) A voluntary federation of organizations in the United States, concerned with the development of standard testing methods.

articulation index (AI) A rating of speech intelligibility, which takes into account both the strength of the talker's voice and the strength of interfering noise. Perfectly intelligible speech has an AI of 1.0, and speech that is made completely unintelligible by interfering noise or weak voice strength has an AI of 0. In the case of speech overheard by occupants of a neighboring office, an AI of 0 indicates complete privacy, and speech that is easily understood in the neighboring office indicates no privacy, or an AI of 1.0.

A-weighted decibel Sound level filtered to a similar frequency sensitivity to that of the human hearing mechanism at normal speech levels; units are in dBA and frequencies below 500 Hz are discounted progressively.

background sound level The sound level in an area, including all sound energy except a specific sound source of interest.

barrier A wall that does not seal with a structural ceiling, floor, or side wall, open to an air space on at least one side.

C-weighted decibel Sound level filtered to a similar frequency sensitivity to that of the human hearing mechanism at high sound levels; units are in dBC and discount low-frequency contributions below 100 Hz slightly.

dead space A room with a very low reverberation time.

dead spot A location in a room where sound, subjectively, appears weak or lacking in reverberation.

decibel (dB) A measure of the strength of a sound field on a logarithmic scale. It may be used to designate the magnitude of the sound level at a point in a sound field or the total sound power level of a sound source. It is defined mathematically as 10 multiplied by the logarithm of the quantity being measured divided by a reference value of the same quantity, where the quantity is related to the power of the source.

diaphragmatic absorber   A light (usually) wood board covering a large air space, useful for low-frequency absorption.

diffraction   The act of sound energy bending around edges of barriers to degrade and limit their sound reduction effectiveness to 10 to 15 dBA.

diffuse field   The area within a room in which sound pressure levels do not vary significantly from place to place, caused by reverberation and diffusion of sound from room surfaces.

diffusion   The scattering of the sound energy in a sound wave after it encounters an irregular or convex reflective surface.

echo   The auditory sensation of hearing, from a single sound source, more than one distinct sound delayed in time, caused by reflected sound paths that differ by more than 100 ft (30 m) from each other in air.

floating floor   A floor that is structurally isolated from a building by using resilient pads and springs between the floor and all rigid connections with the building's structure.

frequency   The rate of oscillation of a sound wave, in units of cycles per second, or hertz (Hz).

Helmholtz resonator   A device having a narrow neck opening into a larger cavity, useful for low-frequency absorption.

hot spot   A location in a room where sound, subjectively, appears stronger than that in surrounding regions, usually resulting from reflections from concave reflective surfaces.

Impact Insulation Class (IIC)   A single-number rating system (defined in ASTM Standard E989) for the sound reduction effectiveness of a floor-ceiling assembly for impact sounds, such as footfalls.

impulsive sound   A sound event that, in a fraction of a second, has a sudden onset and ending.

infrasound   Frequencies below the human hearing sensitivity limits, or below 20 Hz.

insertion loss (IL)   The reduction in sound pressure level after the insertion of a device between a sound source and a listener, usually used to rate the effectiveness of noise barriers and mufflers.

insulation   The ability of a material to reduce airborne sound transmission.

isolation   The ability of a material to reduce structure-borne sound and vibration transmission.

line source   A sound source, such as a train or a stream of vehicular traffic, that has a constant sound output along a line; its propagation path is cylindrical.

live space   A room having a significant amount of reverberation.

masking   Adding acceptable sound to an environment to make unwanted sound inaudible or less annoying.

mass law   Reference to the theoretical fact that a doubling of mass of a homogeneous partition or a doubling of frequency results in a 6-dB increase in transmission loss. In practice, this is only valid over a limited frequency range for any partition.

noise   Unwanted sound.

noise criterion (NC) curves   Curves of noise measured in octave bands and plotted as a function of frequency, which are intended to indicate acceptable (or unacceptable) ambient levels for persons located in particular indoor environments. Updated versions of these curves are known as NCB (balanced noise criterion), which are NC curves extended to include the 31.5-Hz and 63-Hz octave bands, and RC (room criterion) curves, which are straight lines that decrease with increasing frequency at a rate of 5 dB per octave, taking into consideration low-frequency rumble noise.

noise reduction (NR) The reduction in sound pressure level by a partition between rooms, taking into account the transmission loss and the absorption in the room benefited by the partition.

Noise Reduction Coefficient (NRC) The arithmetic average of absorption coefficients in the 250-, 500-, 1000-, and 2000-Hz octave bands (rounded to the closest 0.05); used as a single-number rating for a material's absorption in the human speech frequency range.

octave bands Standard frequency designations (defined in ANSI Standard S1.6) that divide the frequency spectrum into regions that are one octave wide, designated by geometric midfrequencies that are doublings or halvings of 1000 Hz (e.g., 250, 500, or 2000 Hz). In the measurement of sound fields, use of such bands permits the determination of the noise spectrum as a function of frequency rather than as a single number such as dBA or dBC.

point source A sound source that is small in size compared to the distance to a measurement location; its propagation path is spherical.

pure tone A sound signal with its energy at a single frequency.

reflection The act of sound energy bouncing off a surface at the same angle with respect to the surface that the incident wave had.

refraction The changing of direction of sound wave travel caused by changes in medium conditions, such as changes in temperature or density.

resonance   The physical phenomenon by which a material object vibrates at an amplified level when exposed to a specific frequency of sound energy. For example, an organ pipe of a particular length will resonate at its particular resonant frequency when coupled to the source that generates that frequency.

reverberant field   Same as diffuse field.

reverberation   The buildup of sound energy in a room as a result of repeated reflections of sound waves off all room surfaces.

reverberation time (RT_{60})   The time it takes for the sound level within a room to decrease by 60 dB after a loud sound source is turned off; used as a measure of the appropriateness of a room for its purpose.

room resonance   The generation of standing waves in a room at specific frequencies associated with the dimensions of the room.

shadow zone   The area, behind a barrier or below a refracted sound wave, in which sound pressure levels are reduced.

sound concentration   The focusing of sound as the result of reflections off concave surfaces.

sound power level (L_W)   The total sound power of a source expressed in decibels, independent of location with respect to the source; mathematically defined as 10 multiplied by the logarithm of the ratio of the total power radiated in all directions to a standard reference power.

sound pressure level (SPL or L_p)   The decibel level measured at a particular location in space, mathematically described as 20 multiplied by the logarithm of the ratio of the measured acoustic pressure to the pressure associated with the human threshold of hearing. In free space, SPL decreases with increasing distance from a sound source.

sound transmission class (STC) A single-number rating system (defined in ASTM Standard E413) for the transmission loss effectiveness of a partition, based on matching the TL spectrum to a standard curve; STC is appropriate for rating partitions in the human speech frequency range.

spectrum A graphical representation of sound level versus frequency.

transmission coefficient For a sound wave incident on a partition, the ratio of the energy transmitted through the partition to that incident upon the surface. This ratio is usually denoted by the Greek letter τ; τ ranges between 0 and 1, with 0 indicating no transmission and 1 indicating total transmission (as for an open door or window).

transmission loss (TL) A rating, in decibels, of the sound reduction effectiveness of a partition; mathematically defined as 10 multiplied by the logarithm of the reciprocal of (1 divided by) the transmission coefficient.

ultrasound Frequencies above the human hearing sensitivity limits, or above 20,000 Hz.

wavelength The distance between repeating sections of a pure tone sound wave.

American Resources in Acoustics

1. Research/Professional Associations in Acoustics

   • Acoustical Society of America
     Two Huntington Quadrangle
     Melville, NY 11747-4502
     (516) 576-2360
     asa.aip.org

   • Institute of Noise Control Engineering
     P.O. Box 220
     Saddle River, NJ 07458
     (201) 760-1101
     ince.org

2. American Consultants in Acoustics

   • National Council of Acoustical Consultants
     66 Morris Avenue, Suite 1-A
     Springfield, NJ 07081-1409
     (973) 564-5859
     www.ncac.com

- Institute of Noise Control Engineering (board-certified noise control engineers)
  P.O. Box 220
  Saddle River, NJ 07458
  (201) 760-1101
  ince.org

3. American Standards in Acoustics

- American National Standards Institute
  1819 L Street, NW
  Washington, DC 20036
  (202) 293-8020
  www.ansi.org

- American Society for Testing and Materials
  100 Barr Harbor Drive
  West Conshohocken, PA 19428-2959
  (610) 832-9500
  www.astm.org

4. Manufacturers of Acoustical Products

   Key to types of products offered:

   a. Sound-absorptive wall panels

   b. Sound-absorptive ceilings

   c. Spray-on absorptive materials

   d. Fabrics/banners

   e. Diffusive panels

   f. Perforated metal panels

   g. Floating floors

   h. Helmholtz resonator absorbers

   i. Noise control panels/baffles

   j. Enclosures

   k. Windows

   l. Doors

   m. Silencers

n. Lagging/duct wraps

o. Active noise control systems

p. Vibration control products

q. Electronic masking systems

- Acoustic Systems (j)
  415 East Saint Elmo Road
  Austin, TX 78745
  (512) 444-1961
  www.acousticsystems.com

- Armstrong World Industries (a, b)
  2500 Columbia Avenue
  Lancaster, PA 17603
  (717) 397-0611
  www.armstrong.com

- Capaul Architectural Acoustics (a, b)
  1415 Pilgrim Road
  Plymouth, WI 53073
  (800) 876-5884
  www.capaul.com

- Celotex Corporation (b)
  4010 Boy Scout Boulevard
  Tampa, FL 33607
  (813) 873-4212
  www.celotex.com

- Conwed Designscape (a, b)
  800 Gustafson Road
  Ladysmith, WI 54848
  (715) 532-5548
  www.conweddesignscape.com

- Decoustics (a, b)
  65 Disco Road
  Toronto, Ontario, Canada M9W 1M2
  www.decoustics.com

- E-A-R Specialty Composites (i, n, p)
  7911 Zionsville Road
  Indianapolis, IN 46268
  (317) 692-1111
  www.earsc.com

- Eckel Industries Inc. (i, j, m)
  155 Fawcett Street
  Cambridge, MA 02359
  (617) 491-3221
  www.eckelacoustic.com

- Ecophon Certainteed, Inc. (b)
  123 Keystone Drive
  Montgomeryville, PA 18936
  (215) 619-2818
  www.certainteed.com

- Empire International, Inc. (a, f, i, j)
  36744 Constitution Drive
  Trinidad, CO 81082
  (719) 846-2300
  www.empireacoustical.com

- Essi Acoustical Products (a, b, d, i)
  11750 Berea Road
  P.O. Box 643
  Cleveland, OH 44111
  (216) 251-7888
  www.essiacoustical.com

- illbruck-SONEX (a, b, i)
  3800 Washington Avenue North
  Minneapolis, MN 55412
  (612) 520-3620
  www.illbruck-sonex.com

- Industrial Acoustics Company (a, b, i, j, k, l, m)
  1160 Commerce Avenue
  Bronx, NY 10462
  (718) 931-8000
  www.industrialacoustics.com

- Industrial Noise Control, Inc. (a, b, i, j)
  1411 Jeffrey Drive
  Addison, IL 60101
  (630) 620-1998
  www.industrialnoisecontrol.com

- Integrated Interiors Inc. (i)
  21221 Hoover Road
  Warren, MI 48089
  (810) 756-4840
  www.integratedinteriors.com

- Interfinish (f)
  4849 S. Austin Avenue
  Chicago, IL 60638-1492
  (800) 560-5758
  www.interfinish.com

- International Cellulose Corporation (c)
  12315 Robin Boulevard
  Houston, TX 77245-0006
  (713) 433-6701
  www.spray-on.com

- Jamison Door Company (l)
  P.O. Box 70
  Hagerstown, MD 21741
  (800) 532-3667
  www.jamison-door.com

- Kinetics Noise Control (a, g, i, p)
  6300 Irelan Place
  Dublin, OH 43017
  (614) 889-0480
  www.kineticsnoise.com

- Krieger Steel Products Company (k, l)
  4880 Gregg Road
  Pico Rivera, CA 90660
  (562) 695-0645
  www.kriegersteel.com

- Martin Acoustical Products (a, b, i)
  3790 Satellite Boulevard, Suite 100
  Duluth, GA 30096
  (800) 262-7231
  www.martinacoustical.com

- Mason Industries, Inc. (g, p)
  350 Rabro Drive
  Hauppauge, NY 11788
  (516) 348-0282
  www.mason-ind.com

- MBI Products Company, Inc. (a, b, d, i)
  5309 Hamilton Avenue
  Cleveland, OH 44114-3909
  (216) 431-6400
  www.mbiproducts.com

- NCT Group, Inc. (o)
  One Dock Street, Suite 300
  Stamford, CT 06902
  (203) 961-0500
  www.nct-active.com

- Noise Control Systems (a, b, f, j, m, n)
  703 W. 26th Avenue
  Covington, LA 70433
  (504) 892-7973
  www.noisecontrolsystems.com

- Overly Manufacturing Company (l)
  P.O. Box 70
  Greensburg, PA 15601
  (724) 834-7300
  www.overly.com

- The Proudfoot Company, Inc. (a, h, i, n, q)
  Box 276
  Monroe, CT 06468-0276
  (203) 459-0031
  www.thomasregister.com/olc/proudfoot/

- Pyrok, Inc. (c)
  121 Sunset Road
  Mamaroneck, NY 10543
  (914) 777-7070
  www.pyrokinc.com

- Quilite International (a, i)
  8616 La Tijera Boulevard, Suite 509
  Los Angeles, CA 90045
  (310) 641-7701
  www.quilite.com

- RPG Diffusor Systems, Inc. (a, b, e)
  651-C Commerce Drive
  Upper Marlboro, MD 20774
  (301) 249-0044
  www.rpgdiffusors.com

- Saflex/Solutia Inc. (k)
  575 Maryville Centre Drive
  St. Louis, MO 63141
  (314) 674-1000
  www.solutia.com

- The Soundcoat Company, Inc. (i, n, p)
  One Burt Drive
  Deer Park, NY 11729
  (516) 242-2200
  www.soundcoat.com

- Steel Ceilings, Inc. (f)
  P.O. Box 547
  500 N. Third Street
  Coshocton, OH 43812
  (740) 622-4655
  www.steelceilings.com

- Whisper Walls (a, b)
  10957 E. Bethany Drive
  Aurora, CO 80014
  (303) 671-6696
  www.whisperwalls.com

Index

About the Author

Acentech is a multidisciplinary acoustical consulting firm more than 50 years old, with offices in Cambridge, Massachusetts, and Thousand Oaks, California. Acentech senior consultant **James P. Cowan,** principal author of this book, has extensive experience in architectural acoustics and environmental noise issues and has managed hundreds of projects across the country. He conducts seminars and teaches classes on acoustical topics for the American Institute of Architects and other associations, municipalities, educational institutions, and firms. He is a board-certified noise control engineer and the author of *Handbook of Environmental Acoustics, Architectural Acoustics* (a CD-ROM) and many educational articles on acoustical topics.